Interdisciplinary Systems Research
Analysis — Modeling — Simulation

The system science has been developed from several scientific fields: control and communication theory, model theory and computer science. Nowadays it fulfills the requirements which Norbert Wiener formulated originally for cybernetics; and were not feasible at his time, because of insufficient development of computer science in the past.
Research and practical application of system science involve works of specialists of system science as well as of those from various fields of application. Up to now, the efficiency of this co-operation has been proved in many theoretical and practical works.
The series 'Interdisciplinary Systems Research' is intended to be a source of information for university students and scientists involved in theoretical and applied systems research. The reader shall be informed about the most advanced state of the art in research, application, lecturing and metatheoretical criticism in this area. It is also intended to enlarge this area by including diverse mathematical modeling procedures developed in many decades for the description and optimization of systems.
In contrast to the former tradition, which restricted the theoretical control and computer science to mathematicians, physicists and engineers, the present series emphasizes the interdisciplinarity which system science has reached until now, and which tends to expand. City and regional planners, psychologists, physiologists, economists, ecologists, food scientists, sociologists. political scientists, lawyers, pedagogues, philologists, managers, diplomats, military scientists and other specialists are increasingly confronted or even charged with problems of system science.
The ISR series will contain research reports — including PhD-theses — lecture notes, readers for lectures and proceedings of scientific symposia. The use of less expensive printing methods is provided to assure that the authors' results may be offered for discussion in the shortest time to a broad, interested community. In order to assure the reproducibility of the published results the coding lists of the used programs should be included in reports about computer simulation.
The international character of this series is intended to be accomplished by including reports in German, English and French, both from universities and research centers in the whole world. To assure this goal, the editors' board will be composed of representatives of the different countries and areas of interest.

Interdisziplinäre Systemforschung
Analyse — Formalisierung — Simulation

Die Systemwissenschaft hat sich aus der Verbindung mehrerer Wissenschaftszweige entwickelt: der Regelungs- und Steuerungstheorie, der Kommunikationswissenschaft, der Modelltheorie und der Informatik. Sie erfüllt heute das Programm, das Norbert Wiener mit seiner Definition von Kybernetik ursprünglich vorgelegt hat und dessen Durchführung zu seiner Zeit durch die noch ungenügend entwickelte Computerwissenschaft stark eingeschränkt war.
Die Forschung und die praktische Anwendung der Systemwissenschaft bezieht heute sowohl die Fachleute der Systemwissenschaft als auch die Spezialisten der Anwendungsgebiete ein. In vielen Bereichen hat sich diese Zusammenarbeit mittlerweile bewährt.
Die Reihe «Interdisziplinäre Systemforschung» setzt sich zum Ziel, dem Studenten, dem Theoretiker und dem Praktiker über den neuesten Stand aus Lehre und Forschung, aus der Anwendung und der metatheoretischen Kritik dieser Wissenschaft zu berichten.
Dieser Rahmen soll noch insofern erweitert werden, als die Reihe in ihren Publikationen die mathematischen Modellierungsverfahren mit einbezieht, die in verschiedensten Wissenschaften in vielen Jahrzehnten zur Beschreibung und Optimierung von Systemen erarbeitet wurden.
Entgegen der früheren Tradition, in der die theoretische Regelungs- und Computerwissenschaft auf den Kreis der Mathematiker, Physiker und Ingenieure beschränkt war, liegt die Betonung dieser Reihe auf der Interdisziplinarität, die die Systemwissenschaft mittlerweile erreicht hat und weiter anstrebt. Stadt- und Regionalplaner, Psychologen, Physiologen, Betriebswirte, Volkswirtschafter, Ökologen, Ernährungswissenschafter, Soziologen, Politologen, Juristen, Pädagogen, Manager, Diplomaten, Militärwissenschafter und andere Fachleute sehen sich zunehmend mit Aufgaben der Systemforschung konfrontiert oder sogar beauftragt.
Die ISR-Reihe wird Forschungsberichte — einschliesslich Dissertationen —, Vorlesungsskripten, Readers zu Vorlesungen und Tagungsberichte enthalten. Die Verwendung wenig aufwendiger Herstellungsverfahren soll dazu dienen, die Ergebnisse der Autoren in kürzester Frist einer möglichst breiten, interessierten Öffentlichkeit zur Diskussion zu stellen. Um auch die Reproduzierbarkeit der Ergebnisse zu gewährleisten, werden in Berichten über Arbeiten mit dem Computer wenn immer möglich auch die Befehlslisten im Anhang mitgedruckt.
Der internationale Charakter der Reihe soll durch die Aufnahme von Arbeiten in Deutsch, Englisch und Französisch aus Hochschulen und Forschungszentren aus aller

ISR 34

Interdisciplinary Systems Research
Interdisziplinäre Systemforschung

R. H. Atkin

Combinatorial Connectivities in Social Systems

An Application of Simplicial Complex Structures to the Study of Large Organizations

1977 Springer Basel AG

CIP-Kurztitelaufnahme der Deutschen Bibliothek

Atkin, Ronald H.
Combinatorial connectivities in social systems:
an application of simplicial complex structures to the study of large organizations. —
1. Aufl. — Basel, Stuttgart: Birkhäuser, 1977.
(Interdisciplinary systems research; 34)

ISBN 978-3-7643-0912-1 ISBN 978-3-0348-5750-5 (eBook)
DOI 10.1007/978-3-0348-5750-5

Originally published by Birkhäuser Verlag Basel 1977

Preface

This book is a presentation of my recent research work into the theory and applications of mathematical relations between finite sets. The consequent results and concepts are expressed in 'conjunction' with the idea of a union of simplicial complexes and the algorithms and overall methodology have become known as Q-Analysis. The mathematical ideas are essentially combinatorial and are developed in Section-A of the book, whilst Section-B contains an application to a community study (that of the University of Essex).

All of the work has been published elsewhere, in the usual form of papers, and a comprehensive introduction to the whole approach will be found in my earlier book, Mathematical Structure in Human Affairs (Heinemann, London 1974).

Much of the research work referred to in this present book would not have been possible without the co-operation of my research team, of whom I would particularly like to thank J.H. Johnson, and its publication owes a lot to the secretarial help provided by Mary Low.

R.H. Atkin
University of Essex
U.K.

August 1976

Contents

Contents

SECTION B Application to a Community Study

(11)

Contents

8.0 Introduction

8.2 Questions and variables

8.5 More language questions

A0.0 Introduction

This book is the result of recent researches into the mathematical structures associated with relations (as opposed to functions or mappings) and with their applications to communities/town planning/design/systems studies. It is therefore an onslaught on the problem of how to make the "soft sciences" into hard sciences. To this end we must therefore decide on a definition of "hard science" and, having done so, pursue the logic wherever it may lead. As a first consequence we must expect a new insight into what constitutes "data" in a hard science - and, by implication, what errors we must try to avoid in those disciplines (like sociology, psychology, political theory, economics (?), etc.) which are commonly referred to as "soft sciences". We therefore need to distinguish between soft and hard data, and to do so in such a way as to make it amenable to mathematical analysis.

It is here proposed that scientific data is the result of observing set-membership. This kind of data will be referred to as "hard data", whence (by default) all other observations purporting to be "scientific" will be called "soft data". But the idea of set-membership is a non-trivial constraint on the acceptable methodology and is more fundamental than the earlier Kelvin view that "data is what is measurable". Here we are effectively saying that "measuring" is equivalent to the identification of set-membership. In effect we are replacing Kelvin's idea by the following statement, "the only acceptable data is hard data".

But collecting hard data presupposes in the first instance that the set is well-defined. By this we mean that, given a set X, it is possible to say of some proposed element, either that it is, or that it is not, a member of X. Such decisions are binary, being Yes/No or 1/0 type. In the physical laboratory it is common practice to delegate these binary decisions to the physical instruments themselves, thus removing the day-to-day problems from the scientist.

Thus an ammeter says Yes/No to physical candidates for membership of the set of electrical charges. It not only identifies set-members (hard data) but also counts them. This kind of hard data collection is now so commonplace in our technological world that perhaps it induces us to regard the data problem as trivially easy to master. But in the medieval pre-Galilean study of motions of bodies (with its idea that "motion is the realisation of a body's potential") the distinction between soft and hard data was far from obvious and the resolution of the problem occupied many generations of scholars before the decisive break-through was achieved by Galileo. For example, to get some feeling of what the statement about "motion" and "realisation of potential" might have meant we need only move it into our own times and sideways into another (soft) science, whence it could easily become

"education is the realisation of the individual's potential".

To make this into a scientific statement, as hard for future generations as is classical kinetics to this one, we need a mathematical language which describes "education" in terms of relevant hard data. Such a description would have to provide a 'kinematics' for that discipline, comparable to the kinematics of the motion of bodies.

Whereas a geographer can probably feel quite happy about the hardness of much of his data (for example, the set of streets in a town, or the set of contours over a region) there will be situations where he might be tempted to substitute soft data for hard, particularly in those areas where "human geography" creeps over into sociology etc.. In the field of urban studies how do we obtain hard data relevant to the planners asking the question "what urban structure is most conducive to making your town a pleasant place to live?"? Here we need a set whose members are well-defined things called "urban structures" and then we need a set whose members constitute well-defined things called "pleasant places to live". The latter seems to lead inevitably to the question of "value judgements" and where is the hard data in that context? Many social scientists would probably argue that the presence of such judgements must mean that mathematical-set-data is fundamentally irrelevant and so its pursuit is an illusion. But this view ultimately negates the scientific approach and is not unlike the Aristotelian objections to Galileo's definition of velocity (as distance ÷ time) which, apparently, removed all the "content" (or poetry) from motion - in making it hard.

Where is the hard data in the replies to a questionnaire which asks, "do you think that Choice A is worse-than/as-good-as/better-than Choice B?"? On the face of it this presents the citizen with a well-defined set whose members are three in number, viz., worse-than, as-good-as, and better-than. But is this set well-defined among the observers (all of them) qua citizens? Can any two people agree as to worse-than or as-good-as? Also is it clear that any one choice is static, is it not often the case that one feels the need to answer "both worse-than and better-than"? In other words one might really need to make a selection from the power-set, $P(X)$, of such a set X - selecting a subset, rather than a single element - or even from the power set of the power set, $P^2(X)$.

So, well-defined sets certainly require (i) a common agreement about the sets (agreements among the scientific practitioners), and (ii) an appreciation of the logical (and therefore methodological) distinction between a set X and its power set $P(X)$ and so of $P^2(X)$, $P^3(X)$, etc.). This latter point is also fundamental to the avoidance of the many set-theoretical paradoxes which logicians have recently brought to light. A striking instance of this was provided by Bertrand Russell in his famous Barber Paradox, which runs as follows.

"In a certain town (Seville?) every man either shaves himself or he is shaved by the (male) barber. Does the barber shave himself?"

If we assume that the barber shaves himself then we can deduce that he is a man who shaves himself and therefore he is not shaved by the barber. If we assume that the barber does not shave himself then we deduce that he is a man who does not shave himself and therefore that he is shaved by the barber. We therefore obtain a logical paradox - which is illustrated by trying to list a relation λ between two sets Y and X. The set Y = {barber} = {B} whilst the set X = {men} = $\{M_1, M_2, \ldots\}$ and the problem arises when we try to include B in the set X itself. Thus, let λ be a simple matrix of 0's and 1's as shown.

λ	M_1	M_2	M_3	M_4	M_5	M_6	M_7	B
B	1	0	1	1	0	1	0	?

In this scheme the men M_1, M_3, M_4, M_6 are shaved by B, but if B ε X (if the barber is a man) then we cannot decide on the entry under B in the top row. This set X = {M, B} is not well-defined in terms of λ. The reason is that the barber, B, is not a member of X but is in fact a member of P(X). The barber, B, qua barber, is really the subset of X, viz., $\{M_1, M_3, M_4, M_6\}$ ε P(X), and because of that we cannot ask questions of B (of members of P(X)) as if B were a man M_i (a member of X). This distinction, which is logically profound and of fundamental significance in any scientific methodology based on hard data, was classified by Russell in his Theory of Types - in which he insisted that we must not confuse

"elements" with "sets of elements" with "sets of sets of elements" with ...
or X with P(X) with $P^2(X)$ with ...

This suggests that the methodology we are seeking must depend on a hierarchical arrangement of data corresponding to these types, and the hierarchy, H, which I propose is one based on the quite general notion of mathematical cover sets. We can denote the hierarchical levels by N, (N+1), (N+2), etc. (or equally by (N-2), (N-1), N, (N+1), etc.) with possible sets as follows:

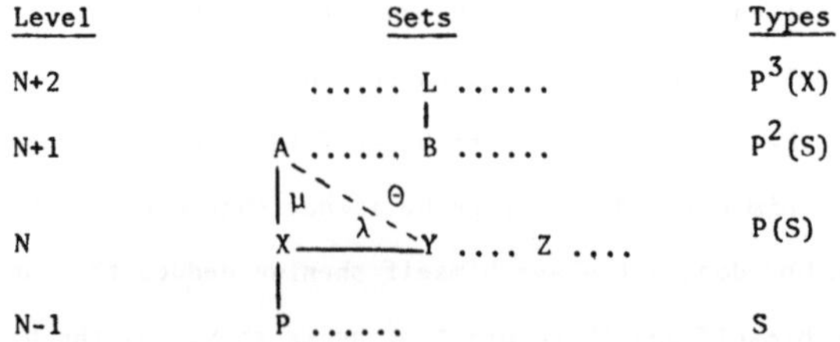

We shall expect the hierarchy of data sets, H, to be defined by mathematical relations λ, μ, ... in the following way.

If X is an N-level set and A a corresponding (N+1)-level set then A must be a <u>cover set</u> for X, that is to say, the elements of A are subsets of X and if

$$A = \{A_1, A_2, \ldots, A_n\}, \quad X = \{X_1, X_2, \ldots, X_m\}$$

then (i) $A_i \in P(X)$ for $i = 1, \ldots, n$

(ii) $X = \bigcup_i A_i$

If, in addition, we know that $A_i \cap A_j = \phi$ (the empty set) then of course A is a <u>partition</u> of X - the elements of A are quite distinct.

The hierarchy H is now defined by relations, like μ, by simply noticing that $A_i \mu X_j$ iff $X_j \in A_i$. This relation μ will therefore be represented by an <u>incidence matrix</u>, M, with binary entries 0/1. Naturally, the whole schema of data will contain "horizontal" relations, like λ, as well as the obvious hierarchical (vertical) relations μ, the two being necessary to allow for such diagonal (joint) relations Θ (where $\Theta = \mu \circ \lambda$).

It follows from the above that the analysis of hard data (suitably arranged in a hierarchical pattern H) depends upon anyanalysis of a <u>relation</u> $\lambda \subset Y \times X$. Such a relation λ, albeit between finite sets, represents a <u>simplicial complex</u> (ref. A-1), denoted by $K_Y(X)$, and this can be intuitively appreciated as a collection of abstract convex polyhedra in a suitable euclidean space E^N. We illustrate this with the following simplex relation λ.

λ	X_1	X_2	X_3	X_4	X_5	X_6	X_7	X_8	X_9	X_{10}
Y_1	1	1	1	1	0	0	0	0	0	0
Y_2	0	1	1	0	1	0	0	0	0	0
Y_3	0	1	0	0	1	1	0	0	0	0
Y_4	0	0	0	0	1	1	1	1	0	0
Y_5	0	0	0	0	0	0	1	1	1	0
Y_6	0	0	0	0	0	0	0	1	1	1
Y_7	0	0	0	1	0	0	0	1	0	1
Y_8	0	0	1	1	0	0	0	0	0	1

In KY(X) the X_i are the vertices; Y_i is the name of the polyhedron (tetrahedron) whose vertices are $\langle X_1 \; X_2 \; X_3 \; X_4 \rangle$; Y_2 is the name of the polyhedron (triangle) whose vertices are $\langle X_2, X_3, X_5 \rangle$; etc.. The polyhedron (the Y_i) are variously connected - we say that (e.g.) Y_1 and Y_2 are 1-connected because they share 2 vertices (these 2 vertices define a common edge, which is a 1-dimensional connection). Let us agree to say that Y_i and Y_j are q-connected if they share (q+1) vertices. Each polyhedron (representing a q-simplex in KY(X)) which possesses (q+1) vertices requires a q-dimensional subspace of E^N for its representation. If n = max (q-values in KY(X)) then n is called the dimension of KY(X) and written n = dim K. There is a well-known mathematical theorem to the effect that it is sufficient to take $N = 2n + 1$ to accommodate the whole of KY(X) in E^N, (v. section A1.2).

Naturally the transpose of the incidence matrix of λ gives a conjugate complex KX(Y), in which the set Y provides the vertices and X provides the names of the simplices (the polyhedra in E^M). A geometrical representation of KY(X) for a relation like λ is given in section A1.2. By defining a suitable transitive relation β, represented by the open statement "—— is q-connected with ——", we shall obtain an equivalence relation on the Y_i (of KY(X)) and so find the equivalence classes of connected components at each q-level. The numbers Q_q of these may be listed and the list of all Q_q numbers (the result of what I have called a Q-Analysis) is a vector, the Structure Vector, for the complex.

All this suggests that a typical hierarchy H of data sets corresponds to a general structure which is the union of complexes like KY(X) and KX(Y), for all the relations λ, μ, ... in H. There is thus a structure (which is well-defined) for each level N, (N+1), ... in H. We represent the structure at N-level by S(N) and refer to it as the static backcloth S(N) for the hard data - the word "static" being clearly relative to the data. Against such a backcloth S(N) the "dynamics" associated with the system must be found. This will consist of "traffic" (of all things which vary in time periods which are short relative to changes in S(N) itself) - except that we must specify

(i) traffic on S(N) ($\equiv$ KY(X)) is determined by the vertex set X

and (ii) traffic can be represented by a pattern on S(N), which is a mapping $\pi : Y \to J$ (J being some number system).

Studies of patterns π and how changes $\delta\pi$ can be interpreted in terms of structural forces (or stresses) have been discussed elsewhere (ref. A-1), but the basic idea rests upon the analogy with an Einsteinian view of the role of the multi-dimensional kinematics. In the first place we notice that a pattern is naturally graded via $\pi = \pi^o + \pi^1 + \ldots \pi^n$, where π^t is the restriction of π to the t-simplices of the complex. Then if, for example, a change δS occurs in the backcloth structure S it will induce a corresponding change $\delta\pi$ in π (by affecting one or more of the terms π^t). Measurements of volume cannot exist when there is no 3-dimensional elements in the backcloth; or if only some of these elements are removed we shall again have $\delta\pi^3 \neq 0$. By comparison with the motion of bodies, in which velocity must be regarded as a π^1 and a non-zero $\delta\pi^1$ represents the presence of an acceleration, we regard $\delta\pi^t$ as a t-dimensional force in the structure, a t-force. Such a force will be experienced by all t-traffic if the change $\delta\pi^t$ is induced by a distortion, δS, of the backcloth S (which is now playing a role as basic as does E^3 or E^4 in the physicist's view of dynamics). These ideas are developed in succeeding chapters of this book - as well as in the author's general work, ref. A-1.

A1.0 Mathematical Relations

We are concerned with binary relations between finite sets. If λ is such a relation between sets Y and X, where

$$Y = \{Y_i;\ i = 1, 2, \ldots m\} \text{ and } X = \{X_i;\ i = 1, 2, \ldots n\}$$

then we write $Y_i\ \lambda\ X_j$ if Y_i is λ-related to X_j. Alternatively we can regard λ as a subset of the cartesian product $Y \times X$ and in that case we write $\lambda \subset Y \times X$. Compatible with this idea is the notation

$$(Y_i, X_j) \in \lambda \quad \text{whenever } Y_i\ \lambda\ X_j.$$

For a given relation $\lambda \subset Y \times X$ there is a naturally defined <u>inverse relation</u> $\lambda^{-1} \subset X \times Y$. This is defined by setting

$$(X_j, Y_i) \in \lambda^{-1} \text{ whenever } (Y_i, X_j) \in \lambda.$$

Each such relation λ may be represented by an <u>incidence matrix</u> Λ in which

$$\Lambda = (\lambda_{ij}) \text{ and } \lambda_{ij} = 1 \text{ if } (Y_i, X_j) \in \lambda$$
$$= 0 \text{ otherwise.}$$

The incidence relation which corresponds to the inverse relation λ^{-1} is then the <u>transpose</u> Λ^T of Λ. Thus $\Lambda^T = (\mu_{ij})$ with $\mu_{ij} = \lambda_{ji}$ for $i = 1, \ldots, m$ and $j = 1, \ldots, n$. It follows that Λ is an $m \times n$ matrix whilst Λ^T is an $n \times m$ matrix.

We also find it convenient, in applications, to consider matrices M which contain entries other than 0 or 1. We assume that the entries are integers $n \in J$, and we shall say that the matrix then corresponds to a <u>weighted relation</u> μ.

Associated with such a μ is a whole series of incidence matrices (binary relations) obtained by the process of slicing, as follows. Take a set of parameters θ_i which are to characterise the matrix M. For example θ_i might refer to row i, for i = 1, ..., m; or θ_i might be θ_{ij}, where each value refers to the (ij)-element of M. Call the $\{\theta_i\}$ the set of slicing parameters for M, and obtain a relation λ by defining its incidence matrix $\Lambda = (\lambda_{ij})$ in terms of elements $m_{ij} \in M$ as follows:

$$\lambda_{ij} = 1 \text{ if } m_{ij} \geq \theta_k \ (m_{ij})$$
$$= \text{otherwise.}$$

Thus, if we "slice by rows" we would select a numerical value for each θ_i, i = 1, ..., m, and take $\lambda_{ij} = 1$ if $m_{ij} \geq \theta_i$ (j = 1, ..., n), zero otherwise. Equally well we can take $m_{ij} \leq \theta_i$ as the condition.

This procedure gives a set of mathematical relations, dependent on the choices of the slicing parameters, and this latter choice is determined by considerations of the applications of the method.

A1.1 Simplicial Complexes

We use the idea of a simplicial complex (or complex of simplices) to represent a mathematical relation $\lambda \subset Y \times X$, in the following way.

If there exists at least one $Y_i \in Y$ such that a (p+1)-subset of X is λ-related to it we call that (p+1)-subset of X a p-simplex. If its members are $\{X_1, X_2, \ldots, X_{p+1}\}$ we write the simplex as

$$Y_i = \langle X_1, X_2, \ldots, X_{p+1}\rangle \text{ or as } Y_i = \sigma_p$$

and call Y_i its name (possibly among many). Any subset of this (p+1)-subset of X is also λ-related to Y_i and is therefore another simplex, say, a q-simplex. This q-simplex, σ_q, is said to be a <u>face of</u> the first σ_p, and this is written

$$\sigma_q < \sigma_p.$$

Each $Y_i \varepsilon \lambda$ therefore identifies a p-simplex (for some p) with all its faces, and this collection of simplices is called a <u>simplicial complex</u> K. More precisely we shall denote the complex by KY(X;λ). This notation is to indicate that the members of the set Y are to be names of simplices whilst the set X is referred to as the <u>vertex set</u> for K.

It is clear that KY(X;λ) is <u>closed</u> under the partial ordering denoted by $<$ and, in our discussion, we shall usually assume that the set Y is the <u>domain</u> of λ whilst X is its <u>range</u>. The <u>vertices</u> of K, or the $X_j \varepsilon X$, can be identified with the 0-simplices (σ_o^i (i = 1, ..., n) and we naturally write

$$\sigma_o^i = \left\langle X_i \right\rangle \text{ for } i = 1, \ldots, n.$$

The largest value of p for which $\sigma_p \varepsilon K$ is called the <u>dimension</u> of K and written dim K. We shall try to reserve the letter N for this quantity, N = dim K.

Otherwise we say that p is the dimension of σ_p (v. next section).

Associated with KY(X;λ) will be that complex defined by λ^{-1}, and this we write as KX(Y; λ^{-1}) and say it is <u>conjugate</u> to the first. When λ is understood we shall refer to these complexes by the symbols KY(X) and KX(Y).

The vertices of the simplex Y_i in KY(X) can be identified by looking along the ith row of the incidence matrix Λ, then X_j is such a vertex if there is a 1 in that j^{th} column. The vertices of the simplex X_j in KX(Y) is Y_i if there is a 1 in the ith row of that j^{th} column. This suggests that much of the computing involved in this study centres around the incidence matrices Λ, Λ^T.

A notation for reminding ourselves of the various roles is the following

λ	X	and	λ^{-1}	Y
Y	KY(X)		X	KX(Y)

Since the empty set is a subset of every set it seems reasonable to include it as a face of every simplex in KY(X). When this is done we denote the empty set by the (-1)-simplex, σ_{-1}, and say that the complex K is thereby augmented; we write it as K^+ or as $K \cup \sigma_{-1}$..

Finally, there is a natural way in which a complex K may be given an orientation, ω. Since the vertices, say X, are given the ordering of the natural numbers we say that the p-simplex

$$\sigma_p = \langle X_{\alpha_1}, X_{\alpha_2}, \ldots, X_{\alpha_{p+1}} \rangle$$

possesses a positive orientation if the sequence

$$\{\alpha_1, \alpha_2, \ldots, \alpha_{p+1}\}$$

is an even permutation of the same numbers with their natural ordering, and that it possesses a negative orientation when that permutation is odd. In the first place we denote the simplex by σ_p, or $+\sigma_p$, and in the second case by $-\sigma_p$. In this way every simplex σ_p, $p > 0$, possesses an orientation which is naturally induced and when this is done we say that the complex K possesses an orientation.

A1.2 A Geometrical Representation

A complex $KY(X;\lambda)$ can be represented in a euclidean space E^H in the following way, for a suitable choice of H.

Each p-simplex, typically $\sigma_p = \langle X_1, X_2, \ldots, X_{p+1}\rangle$, is made to correspond to a convex polyhedron in E^H with (p+1) vertices which themselves correspond to X_1, X_2, X_{p+1}. Thus, in an intuitive sense, in E^H the simplex σ_p is represented by the "solid" polyhedron with (p+1) vertices. The complex KY(X) is then represented by a collection of polyhedra suitably connected to each other by sharing "faces" (or sub-polyhedra).

This geometrical representation is always possibly by suitably choosing H, it being remembered that each σ_p requires p dimensions for its accommodation. It might be somewhat surprising therefore that an economical theorem exists which demonstrates that the value of H need only be taken as

$$H = 2N + 1, \text{ where } N = \dim K.$$

To prove this, let $X = \{X_1, \ldots, X_m\}$ be the vertex set and take the set of points $\{P_1, \ldots, P_m\}$ in E^{2N+1}, where

$$P_r = (r, r^2, \ldots, r^{2N+1}), \quad r = 1, \ldots m$$

(i) Any (2N+2) points of the set $\{P_r\}$ are independent, because otherwise there would exist scalars $C_1, \ldots, C_{2N+2}$ (not all zero) such that

$$C_1 P_{r_1} + \ldots + C_{2N+2} P_{r_{2N+2}} = 0$$

together with the condition, $\Sigma C_r = 0$.
This means that there would be (2N+2) equations for the scalars, viz.,

$$\sum_{1}^{2N+2} C_k = 0, \quad \sum_{1}^{2N+2} C_k\, r_k = 0, \quad \ldots, \quad \sum_{1}^{2N+2} C_k\, r_k^{\,2N+1} = 0$$

but we know that

$$\det(1\ r\ r^2\ \ldots\ r^{2N+1}) = \sum_{p < q} (r_p - r_q) \neq 0$$

and so $C_r = 0$, for $r = 1, \ldots (2N+2)$.

It follows that any subset of these (2N+2) points $\{P_r\}$ is also an independent set.

(ii) Now we map each X_r to P_r, and each p-simplex σ_p to its convex set on (p+1) vertices, via

$$\mu : X_r \to P_r,\ r = 1, \ldots, m$$

and $$\sigma : \langle X_1 X_2 \ldots X_{p+1} \rangle \to \text{convex set on } \{P_1 P_2 \ldots P_{p+1}\}.$$

All the $\sigma_p \in K(X)$ are accommodated in E^{2N+2} if the distinct simplices are disjoint or only meet in a commonface (which is another polyhedron in E^{2N+2}). But a common point P of two p-simplices will be given by

$$P = \Sigma\, a_r P_r = \Sigma\, b_r P_r$$

together with $$\Sigma\, a_r = \Sigma\, b_r = 1,$$

and such that not more than (p+1) of the a's (or the b's) are non-zero. Hence $\Sigma(a_r - b_r)P_r = 0$ together with $\Sigma(a_r - b_r) = 0$, and not more than (2p+2) of the $(a_r - b_r)$'s are non-zero. But $p \leq N$ and so, by (i) above, the $\{P_r\}$ is an independent set and so $a_r = b_r$, for all r. Hence the common points of two simplices lie in a common simplex in E^{2N+2}. This result is independent of how many N-simplices are contained in K, or of how they are mutually connected. But this means that, in the realm of Social Science, the euclidean spaces which can accommodate the various mathematical relations will commonly be much greater than the 3-dimensional structure demanded by the physicist.

We illustrate the geometrical representation by a simple numerical example, and allow for the fact that perspective limits a plane representation to a value of $N = 3$.

Example Given $\lambda \subset Y \times X$, $m = 6$, $n = 8$, and the incidence matrix as

$$\Lambda = \begin{bmatrix} 1 & 1 & 1 & 1 & 0 & 0 & 0 & 0 \\ 0 & 0 & 1 & 1 & 1 & 0 & 0 & 0 \\ 0 & 0 & 0 & 0 & 1 & 0 & 0 & 1 \\ 0 & 0 & 0 & 0 & 0 & 1 & 1 & 1 \\ 0 & 0 & 1 & 0 & 0 & 0 & 1 & 0 \\ 0 & 0 & 0 & 1 & 0 & 1 & 0 & 1 \end{bmatrix}$$

Considering $KY(X;\lambda)$:

$\langle X_1\ X_2\ X_3\ X_4\rangle$ is a σ_3 whose name is Y_1 (row 1)

$\langle X_3\ X_4\ X_5\rangle$ is a σ_2 whose name is Y_2

$\langle X_6\ X_7\ X_8\rangle$ is a σ_2 whose name is Y_4

and similarly, $\langle X_4\ X_6\ X_8\rangle = Y_6$, $\langle X_5\ X_8\rangle = Y_3$, $\langle X_3\ X_7\rangle = Y_5$.

A geometrical representation of $KY(X)$ is given in Figure 1, whilst $KX(Y)$ is shown in Figure 2.

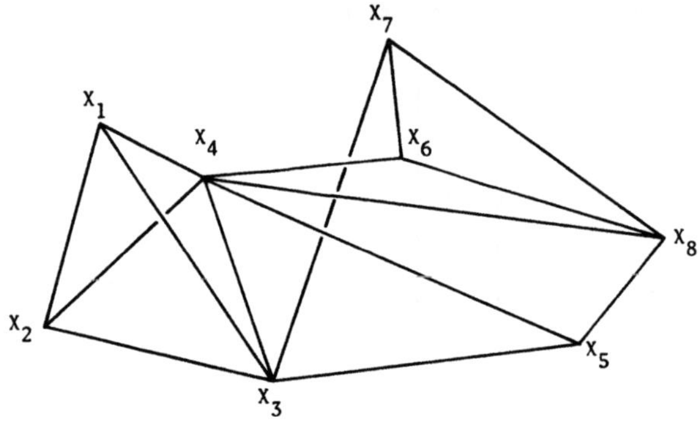

Figure 1 $KY(X;\lambda)$

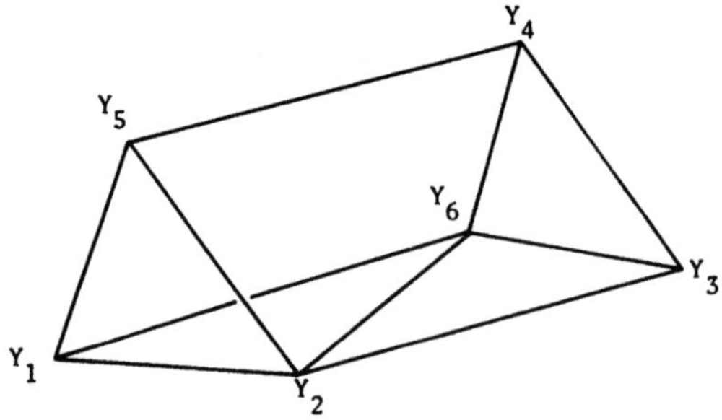

Figure 2 $KX(Y;\lambda^{-1})$

A1.3 Chains of q-connection in K

Given two simplices σ_p, σ_r in K we shall say they are joined by a chain of connection if there exists a finite sequence of simplices

$$\sigma_{\alpha_1}, \sigma_{\alpha_2}, \ldots, \sigma_{\alpha_h}$$

such that (i) $\sigma_{\alpha_1} \leq \sigma_p$

(ii) $\sigma_{\alpha_h} \leq \sigma_r$

(iii) σ_{α_i} and $\sigma_{\alpha_{i+1}}$ share a common face (say)

σ_{β_i}, $i=1, \ldots, (h-1)$

We shall say that this sequence is a <u>chain of q-connection</u> (or that it is a <u>q-connectivity</u>) if q is the <u>least</u> of the integers

$$\alpha_1, \beta_1, \beta_2, \ldots, \beta_{h-1}, \alpha_h$$

The <u>length</u> of the chain will be taken as (h-1) and, when needed, the chain may be denoted by $[\sigma_p, \sigma_r]_q$.

As a special case we see that a p-simplex σ_p must be p-connected to itself, by a chain of length 0, although it cannot be (p+1)-connected to any simplex.

It is easy to see that if σ_p and σ_r are q-connected then they are also (q-1)-, (q-2)-, ..., 1-, 0-connected in K.

If we refer to the numerical example of section A1.2, and Figure 1, we can see that

$$Y_1 = \langle X_1\ X_2\ X_3\ X_4\rangle \text{ is 1-connected to } Y_2 = \langle X_2\ X_4\ X_5\rangle$$

since they share the 1-simplex $\langle X_2\ X_4\rangle$, whereas

$$Y_5 = \langle X_3\ X_7\rangle \text{ is (only) 0-connected to } Y_4 = \langle X_6\ X_7\ X_8\rangle$$

Similarly we can see that

$$Y_1 \overset{0}{—} Y_5 \overset{0}{—} Y_4 \overset{1}{—} Y_6$$

is a 0-connectivity.

For a particular simplex, σ_r, in K it is consequently possible to identify two values of q. The first value is $\check{q}$ (<u>the bottom-q</u>) which is the least value of q for which σ_r is q-connected to any distinct simplex. The second value is $\hat{q}$ (the <u>top-q</u>) which is the dimension of that simplex (in this case $\hat{q} = r$). These two values have been referred to in Research Report I, and elsewhere, where their relevance in applications has been explored. Closely associated with $(\check{q}, \hat{q})$ is a quantity called the <u>eccentricity</u> of σ. This is a rational number given by

$$\text{Ecc }(\sigma) = \frac{\hat{q} - \check{q}}{\check{q} + 1}$$

This quantity is defined for all values except $\check{q} = -1$, and then we say Ecc $(\sigma) = \infty$. This "infinite" value for the eccentricity occurs when σ is totally disconnected from the rest of the complex. Possible significance in its use as a measure is discussed in [2, 3].

A1.4 Q-Analysis

The process of identifying those pieces of K which are q-connected, for all values of q from 0 to N, constitutes a sequence of partitionings of the simplices of K. To see this we introduce, for a fixed q, a relation

γ_q on the simplices of K, defined by

$$(\sigma_p, \sigma_r) \in \gamma_q \text{ if and only if}$$

$$\sigma_p \text{ is q-connected to } \sigma_r$$

This γ_q is reflexive, symmetric and transitive and is therefore an equivalence relation. The equivalence classes, under γ_q, are the members of the quotient set $K/_{\gamma_q}$, and constitute a partition of K (or more accurately of all simplices of K which are of order $\geq$ q).

We denote the cardinality of $K/_{\gamma_q}$ by Q_q. This equals the number of distinct q-connected components in K. When we analyse K by finding all the values of

$$Q_0, Q_1, Q_2, \ldots, Q_N \text{ where } N = \dim K$$

we say that we have performed a Q-analysis on K.

Referring to the numerical example of KY(X;λ) and Figure 1 in section A1.2 we can obtain these Q_q-values by inspection. Since there is only one σ_3, $\langle X_1\ X_2\ X_3\ X_4\rangle$, we have $Q_3 = 1$. At the level of q = 2 the separate components are $\langle X_1\ X_2\ X_3\ X_4\rangle$, $\langle X_3\ X_4\ X_5\rangle$, $\langle X_4\ X_6\ X_8\rangle$ and $\langle X_6\ X_7\ X_8\rangle$, no two of which are 2-connected. Thus $Q_2 = 4$ since now Y_1 and Y_2 share a 1-face, as do Y_4 and Y_6, but Y_3 and Y_5 are new 1-simplices. At q = 0 we have $Q_o = 1$ since the complex K is in "one piece".

The value of Q_o is the same as the topologists zero-order Betti number; it gives the number of (arc wise) connected components of K. The higher order Betti numbers equal the number of free generators of the homology groups. These are not the same as the higher Q-numbers. Our analysis therefore gives a generalisation of the zero order Betti number different from that provided by homology theory. A discussion of the latter showing, for example, that the homology groups of conjugate complexes are isomorphic is given in Dowker [4].

A1.5 An Algorithm for Q-Analysis

If the cardinalities of the sets Y and X are m, n respectively, the incidence matrix Λ is an $(m \times n)$-matrix with entries 0 or 1. In the product $\Lambda\Lambda^T$ the number in position (i, j) is the result of the inner product of row i with row j, taken from Λ. This number equals the number of 1's common to rows i and j in Λ. It is therefore equal to the value (q+1), where q is the dimension of the shared face of the simplices σ_p, σ_r represented by rows i and j. The algorithm can therefore be summarised as follows:

To find the shared face q-value between all pairs of the Y's in KY(X; λ),

(i) form $\Lambda\Lambda^T$, an $(m \times m)$-matrix,

(ii) evaluate $\Lambda\Lambda^T - \Omega$, where $\Omega = (w_{ij})$ and $w_{ij} = 1$ for $i, j = 1, 2, \ldots, m$.

The analysis for KX(Y;λ^{-1}) follows by forming $\Lambda^T\Lambda - \Omega^1$ where Ω^1 is an $(n \times n)$-matrix of 1's.

The complex KY(X;λ) of section 1.2 gives the following q-pattern for the shared faces. We reproduce only the upper triangular "half" of the symmetric matrix $\Lambda\Lambda^T - \Omega$, for ease of analysis, and where we write - for q = 1 (the case of disconnection).

Y_1	Y_2	Y_3	Y_4	Y_5	Y_6	
3	1	-	-	0	0	Y_1
	2	0	-	0	0	Y_2
		1	0	-	0	Y_3
			2	0	1	Y_4
				1	-	Y_5
					2	Y_6

The integers in the diagonal are the dimensions of the Y_i as simplices; the Q-analysis follows by inspection.

$N = 3 = \dim K$ since Y_1 is a 3-simplex

at $q = 3$ we have $Q_3 = 1$; viz., $\{Y_1\}$

$q = 2$ we have $Q_2 = 4$; $\{Y_1\}$, $\{Y_2\}$, $\{Y_4\}$, $\{Y_6\}$

$q = 1$ we have $Q_1 = 4$; $\{Y_1, Y_2\}$, $\{Y_3\}$, $\{Y_4, Y_6\}$, $\{Y_5\}$

$q = 0$ we have $Q_o = 1$; {all}.

A1.6 Q-space and the Structure Vector

We associate with the whole family of complexes a vector space over a field F which we shall call Q-space. (In fact, in our studies so far we are only concerned with the ring of integers J in place of F. This makes the Q-space an algebraic module instead of a vector space). This must be of sufficient dimension to accommodate any of the finite dimensional complexes which interest us. Then, given a complex K, with dim K = N, we introduce a (first) structure vector $\underline{Q}(K)$,

$$\underline{Q} = \{Q_N, Q_{N-1}, \ldots, Q_1, Q_o\}.$$

In the numerical example discussed in section A1.5 we obtain the structure vector

$$\underline{Q} = \{\overset{3}{1} \quad 4 \quad 4 \quad \overset{0}{1}\}$$

where the superscripts denote the q-values.

This Q-space is not necessarily equipped with an inner product, or a metric, but possesses only the structure of the vector space (or module).

If $Q_o > 1$ the complex K is in Q_o pieces and can therefore be regarded as the union of disjoint complexes; the resultant $\underset{\sim}{Q}$ is then the vector sum of the separate $\underset{\sim}{Q}_i$ associated with the K_i, where

$$K = \bigcup_i K_i$$

It later transpires that there is (applied) significance in a vector which is closely associated with $\underset{\sim}{Q}$, and defined by it. This associated vector is given the name of obstruction vector and written $\hat{\underset{\sim}{Q}}(K)$, or simply $\hat{\underset{\sim}{Q}}$.

We restrict ourselves to a complex K for which $Q_o = 1$ and introduce the vector $\underline{U}$ which gives the unit point (affinely), viz.,

$$\underline{U} = \{1, 1, \ldots, 1\}$$

and we shall use the letter $\underline{U}_r$ to denote the projection of $\underline{U}$ on the r-dimensional subspace containing $\underset{\sim}{Q}(K)$. The obstruction vector is now defined as

$$\hat{\underset{\sim}{Q}} = \underset{\sim}{Q} - \underline{U}_{N-1}$$

or

$$\hat{\underset{\sim}{Q}} = \{Q_N, \quad Q_{N-1} - 1, \ldots, Q_1 - 1, Q_o - 1\}$$

where, of course, $Q_o - 1 = 0$.

In later sections we shall see that this obstruction vector $\hat{\underset{\sim}{Q}}$ is a property of the complex K which acts as an obstruction (or obstacle) to the free flow of patterns on K.

In general we can see our structure vector $\underset{\sim}{Q}(K)$ as a first measure of some of the global structure to be found there by virtue of the property of q-connectivity. It overlooks, of course, the internal structure of any one component of K.

A1.7 Patterns on a Complex

By a pattern π, on a complex K, we shall mean a mapping

$$\pi : \{\sigma_p^{\ i},\ 0 \leqslant p \leqslant N,\ \text{all } i\} \rightarrow J$$

where J is (usually) the integers. Thus π is defined on every simplex of K and, because these are graded by their q-values it is natural to grade the pattern itself. Thus we can write

$$\pi = \pi^o \oplus \pi' \oplus \ldots \oplus \pi^t \oplus \ldots \oplus \pi^N$$

where N = dim K, and where

$$\pi^t = \pi \mid \{\sigma_t^{\ i};\ \text{fixed } t\}$$

Each π^t is therefore a set function, defined on specified (t+1)-subsets of the vertex set X of K.

The mapping need not be (1-1) nor, at this stage, is there any question of its being an homomorphism.

The "face" ordering, $\leqslant$, is to be manifest by the restriction map, so that,

$$\sigma_q \leqslant \sigma_p \text{ implies that } \pi(\sigma_q) = \pi \mid \sigma_q.$$

A1.8 Dynamics on a Static Backcloth Structure

We are now in a position to see in detail a more profound role for the abstract geometrical structure which constitutes the static backcloth S(N). This role which we claim for S(N) is comparable to that which the physicist claims for his standard 3-dimensional space - the Euclidean space E^3. This classical rigid space, E^3, which has played such a dominant role in orthodox physical theories since it was consciously introduced by Descartes, and accepted by Gallileo and Newton, is a 3-dimensional structure and exhibits a connectivity pattern - in our terms - which is particularly simple. This connectivity pattern gives a structure vector with a single component at the levels $q = 3, 2, 1, 0$. Set against this particular structure we find the physical scientist proposing the notion of a pattern π in the sense which we have already discussed. In fact he has introduced a whole series of patterns which he has used to represent physical phenomena of various kinds. Perhaps the simplest of these is the pattern he has used to describe the velocity of a particle. The "particle" has been for the physicist the physical manifestation of his notion of a zero order simplex, or a geometrical point. By observing the motion of such a particle he has believed himself to be finding a physical manifestation of a zero order chain in the sense of our analysis. Thus he has identified a sequence of points in his space E^3, which, taken in pairs, he has been able to identify as a chain of one-simplices joined at their ends (thus producing a 0-chain). Now on each one of these 1-simplices, or edges, he has produced a mapping which gives him a number (usually in the real number system). This mapping is what he has described as velocity of the particle - other things being equal. Thus the physicist in talking about velocity has in fact been discussing the notion of a pattern on a particularly simple low-dimensional static backcloth.

The use of this kind of pattern by the physical scientist has provided him with a mathematical way of studying the dynamics of his system. This word "dynamics" can therefore be regarded as synonymous with the word "patterns" - provided these patterns are associated with a suitably observed (or assumed) static backcloth S(N). It has of course been convenient, and intellectually attractive, to regard this static backcloth as static and rigid and permanent. The behaviour of properties of the phenomena which are observed in this physicists' world have therefore all been attributed to the peculiarities of specially defined patterns on S(N). Thus the motion of the planets, as postulated by Gallileo, Kepler, and explained by Newton, has been described in terms of a special pattern (or mapping) on the same static Euclidean backcloth S(N). In this backcloth the Sun becomes a physical manifestation of a single point, or 0-simplex, whilst a pattern such as the Earth becomes another such 0-simplex. Each of them moves over a series of positions (or 0 simplices) and in so doing demonstrate a physical manifestation of 1-simplices, 0-connected to each other. The pattern which is described as the "inverse square law of gravitation" then becomes a dynamic explanation for the phenomena observed. Notice that in such a description the static backcloth remains unaltered and the dynamic properties are attributed to specific dynamic patterns defined on the backcloth.

An alternative view of this kind of phenomena was given by the Einstein formulations of relativity at the beginning of this century. In such a formulation, for example, the observed orbits of the planets become "natural" paths in a geometry which has been suitably altered. This alteration of the geometry amounts to the imposition of special constraints on the static backcloth S(N). This constraint - or pseudo metric, after Einstein - alters the connectivity of the static backcloth. Precisely, we mean that the manifestations of the abstract zero-simplices in S(N) become the observed positions of the particles (or planets) and these observed positions, or paths, are <u>those which the geometry now permits</u>. They are not the paths which the supposed Newtonian forces induce - they are the paths which the geometry permits. Thus the modified geometry has effectively had built into it that structure which had previously been blamed on the specialised patterns which were invented to describe the paths in a rigid and unchanging backcloth.

But the physical scientist has not of course been entirely concerned with the concept of particles and with those of velocity, acceleration, etc.. Many other instances of patterns on a rigid backcloth S(N) have been developed in the course of the subject. Thus the nineteenth century idea, or dedication, that science consists of measuring quantities is in fact an expression of the idea that a pattern (with suitable values in common arithmetic) must be found as the explanation of any scientifically observable phenomena. For example, measuring temperature is the attribution of a pattern on a suitable backcloth. In the same way we interpret the measurement of light intensity, sound, magnetic field, electric charge, etc.. It is the study of changes in these patterns which then naturally lead to an appreciation or interpretation of one's intuitive sense of dynamics.

Now the changes of patterns are an expression of what, in the Newtonian sense, we have traditionally called a force. Thus the change of velocity, what is generally and loosely written as δv, is already a measure of some force (or possibly impulse) which we claim the "particle" is experiencing in the backcloth S(N). It is of course common, particularly in an elementary education, to speak of the force as causing the change in the velocity. But this, on closer examination, tends to lead us in a circular path, so that it is more realistic to regard the change of pattern as a description of the same thing as the word force is a description of. Thus we might as well speak of a change in pattern as defining a force in the structure - that is to say in the static backcloth S(N) wherein we find the phenomena.

We shall accordingly adopt this same convention in this study and we shall speak of a change of pattern as representing, or as being, a force experienced by something or other in the backcloth S(N). It immediately follows of course that, because a pattern is a graded quantity, then the force is a graded quantity and we must in general expect to distinguish between a force which is associated at one q-value and a force associated at another q-value. With this in mind we therefore introduce the following definition.

Definition: If $\delta\pi$ is an incremental change in a pattern π, defined on a backcloth S(N), and if $(\delta\pi)^t \in \delta\pi$ is the t-dimensional (non-zero) component of the graded $\delta\pi$, then we say that it is associated with a t-force in S(N). We call the ratio

$$(\sigma_t, (\delta\pi)^t) \div (\sigma_t, \pi^t)$$

when it exists, the intensity of the t-force on the simplex $\sigma_t \in S(N)$.

Of course when we use this definition we are adopting what we might loosely call the Newtonian view of the dynamics on the backcloth. We shall see, when we look at the details of the specific patterns that the alternative view - what we might as well refer to as the Einsteinian view - is in many ways the more natural. In this latter case we shall consider changes in the geometry which allow free changes in the patterns - and by free we shall mean that the changes are compatible with the geometrical backcloth. But in the former case, when we are using the Newtonian description, there are many instances in which the common usage of the English language seems to be expressing the notions of forces or pressures in the structure. Thus we shall find instances of changes of pattern which are very naturally, and would commonly be, described in terms of social forces or social pressures, or organisational pressures, or ambitions, etc. which are experienced by the individual person who is living his life in that particular structure.

A2.0 Exterior Algebra

In this section we review some definitions and elementary properties of exterior algebra.

We begin with a finite dimensional vector space V which has a basis $\{u_1, u_2, \ldots, u_n\}$. By imposing a "product" on the elements of this basis we can transform V into an algebra (which must possess two binary operations, + and × ,suitably compatible). This product is to be the exterior product, or wedge product, and is commonly denoted by ^; its characteristic is that it is antisymmetric.

Define the wedge product by requiring

$$u_i \wedge u_j = -u_j \wedge u_i \quad \text{(antisymmetry)}$$

and that ^ is distributed over +. Then we have, for example,

$$u_i \wedge u_i = 0 \quad \text{(nilpotence)}$$

and so $(a_1u_1 + a_2u_2) \wedge (b_1u_1 + b_2u_2) = (a_1b_2 - a_2b_1)\, u_1 \wedge u_2$

This wedge product gives us a new vector space spanned by all the products $\{u_i \wedge u_j\}$, $i < j$, as a basis. We denote this new vector space by Λ^2V, the superscript 2 denoting the pairs of basis vectors.

In a similar way we can form the vector spaces

$$\Lambda^3V,\ \Lambda^4V,\ \ldots,\ \Lambda^nV$$

whose bases are triples of u_i, etc.. Notice that we cannot form a product containing more than n basis elements u_i because that would mean that one of them appeared twice, leading to zero.

We call the direct sum of these vector spaces "the" wedge product of V and adjoin to the above both V itself (which we write as Λ^1V) and the field of scalars F (which we write as $\Lambda^0 V$). Thus we obtain the total product

$$\Lambda V = \Lambda^0 V \oplus \Lambda^1 V \oplus \Lambda^2 V \oplus \ldots \oplus \Lambda^p V \oplus \ldots \oplus \Lambda^n V$$

Since sum (+) and product ($\wedge$) are well-defined on ΛV, as is multiplication by scalars, we can regard it as an algebra (the exterior algebra on V) provided we ensure that $\wedge$ is distributive over +. Since the product $\wedge$ is associative, that is $(u_i \wedge u_j) \wedge u_k = u_i \wedge (u_j \wedge u_k)$, the algebra ΛV is an associative algebra.

Example Take a 3-dimensional vector space V, over the reals R, with basis $\{u_1, u_2, u_3\}$. Then the exterior algebra $\{\Lambda V, +, \wedge\}$ is graded by

$$\Lambda V = \Lambda^0 V + \Lambda^1 V + \Lambda^2 V + \Lambda^3 V$$

and $\Lambda^0 V = R$. The product of two elements of $\Lambda^1 V$ lies in $\Lambda^2 V$; for example,

$$(a_1 u_1 + a_2 u_2 + a_3 u_3) \wedge (b_1 u_1 + b_2 u_2 + b_3 u_3) \wedge (v_1 u_1 + c_2 u_2 + c_3 u_3)$$

$$= A(a, b, c)\ (u_1 \wedge u_2 \wedge u_3)$$

where the scalar A(a, b, c) is the expansion of the determinant

$$A(a, b, c) = \begin{vmatrix} a_1 & a_2 & a_3 \\ b_1 & b_2 & b_3 \\ c_1 & c_2 & c_3 \end{vmatrix}$$

This determinant A(a, b, c) is an example of what is called a pseudoscalar; it is the one component of a vector in a 1-dimensional space $\Lambda^3 V$, just as $\Lambda^0 V$ is a 1-dimensional space, but its sign depends on the order of the vectors in the basis $u_1 \wedge u_2 \wedge u_3$.

In the general case, when V is an n-dimensional vector space, we notice that the dimension of the space $\Lambda^p V$ is ${}^nC_p$, the number of combinations of n things p at a time. Hence

$$\dim \Lambda^p V = \dim \Lambda^{n-p} V$$

because $$^nC_p = {}^nC_{n-p}$$

The space $\Lambda^{n-1}V$, for this reason, is called the space of pseudovectors.

A general element x in the algebra will be the sum of elements out of the separate vector spaces Λ^pV, thus we would write

$$x = x_o + x_1 + \ldots + x_p + \ldots + x_n$$

where $x_p \in \Lambda^pV$. The product rule then ensures that

$$x_p \wedge x_q \in \Lambda^{p+q}V$$

provided $p + q \leqslant n$.

We notice that if $x \in \Lambda V$ then $x^2 = 0$ (nilpotency) and so if we consider a <u>polynomial</u> in ΛV it can only be of the form

$$a_o + a_1x \qquad a_o, a_1 \in F$$

or, when $x \neq y$, $a_o + a_1x + b_1y + c_2xy \qquad a_o, a_1, b_1, c_2 \in F$.

Since each $x \in \Lambda V$ can be written in terms of its homogeneous components it follows that a polynomial in ΛV (coefficients in F) will always be of the form of

$$a_o + a_1{}^iu_i + a_2{}^{ij}\, u_iu_j + \ldots + a_n\, u_1u_2 \ldots u_n$$

where we have written u_iu_j for $u_i \wedge u_j$.

Thus a polynomial is another way of looking at an element of ΛV, when we wish to emphasise its derivation from a (wedge) product like x y z, etc..

A2.1 An Algebraic representation of K

Let X be totally ordered, by the natural ordering of the integers and let the set

$$V = \{x_1, x_2, \dots, x_n\}$$

be made to correspond (1-1) to the set $X = \{X_1, X_2, \dots, X_n\}$. Construct the exterior algebra ΛV, over the n-dimensional module $\{V, +, J\}$, in which $\Lambda^o V = J$.[4]

Then there is an algebraic representation of K via the map

$$\rho = K \to \Lambda V$$

where (i) $\rho = \{\rho_p;\ p = o, 1, \dots, N\}$

(ii) $\rho_o : \langle X_i \rangle \to x_i \in \Lambda^1 V$

(iii) $\rho_p : \langle X_{\ 1}, X_{\ 2}, \dots, X_{\ p+1} \rangle \to x_{\alpha 1}\, x_{\alpha 2} \dots x_{\alpha p+1} \in \Lambda^{p+1} V$

$p=1, 2, \dots, N$ and $N \leq n-1$

(iv) when K is augmented, there exists a map ρ_{-1} such that

$\rho_{-1} : \sigma_{-1} \to 1 \in J$

This representation means that K possesses an orientation (in ΛV) expressed in the antisymmetry of the exterior product.

Thus

$$\langle X_i\ X_j \rangle = -\langle X_j\ X_i \rangle \quad \text{in K}$$

becomes

$$x_i\ x_j = -x_j\ x_i \quad \text{in } \Lambda V$$

The natural grading of the exterior algebra, viz.,

$$\Lambda V = \Lambda^o V \oplus \Lambda^1 V \oplus \dots \oplus \Lambda^{p+1} V \oplus \dots \oplus \Lambda^{N+1} V \oplus \dots \oplus \Lambda^n V$$

corresponds to the grading of the augmented K, viz.,

$$K = K^{-1} \oplus K^o \oplus \dots \oplus K^p + \dots \oplus K^N$$

in which

$$K^p = \{\sigma_p{}^i;\ \text{fixed p}\}$$

and K^p corresponds to $\Lambda^{p+1}V$

Now every p-simplex in K, σ_p, possesses a unique representation (via products of distinct basis elements) in $\Lambda^{p+1}V$, and so the whole complex K may be represented by a polynomial in ΛV, of the form

$$\pi_o = 1 + x_1 + x_3 + x_1x_2x_4 + x_3x_2x_7 + \text{etc.},$$

representing, in this algebraic sense, the formal sum

$$\pi_o = \sigma_{-1} + \sum_i \sigma_o^{\,i} + \sum_i \sigma_1^{\,i} + \ldots + \sum_i \sigma_p^{\,i} + \ldots + \sum_i \sigma_N^{\,i}.$$

A2.2 Patterns as polynomials in ΛV

If we interpret a monomial term, in ΛV, (say)

$$\alpha\,\sigma_p \text{ with } \alpha \in J$$

as meaning that the value α is attributed to the simplex σ_p, then we see that every pattern π corresponds to the formal sum, which will also be denoted by π, viz.,

$$\pi = \alpha_{-1}\,\sigma_{-1} + \sum_i \alpha^o_{\,i}\,\sigma_o^{\,i} + \ldots + \sum_i \alpha_i^{\,N}\,\sigma_N^{\,i}$$

This is a polynomial, with coefficients in J, in the algebraic representation ΛV. When we wish to distinguish strictly between π as a map and π as a polynomial we shall refer to the latter as a <u>pattern polynomial</u>, π.

The complex K itself, by the result in [2], thereby becomes represented by a pattern π_o in which each $\alpha^p_{\,i} = 0$ or 1, depending whether the simplex $\sigma_p^{\,i}$ is absent/present in K. This pattern π_o, which defines K itself, is an expression of what elsewhere has been called the static backcloth K[2].

Each pattern π is naturally a graded polynomial, being the sum of terms in $\Lambda^p V$, for $p \geqslant 0$, but in order to relate this grading to that of the complex K itself we shall adopt the convention inherent in writing π as follows:

$$\pi = \pi^{-1} + \pi^{o} + \ldots + \pi^{N}$$

when K is augmented (π^{-1} being absent otherwise), with $\pi^p \in \Lambda^{p+1}V$. Thus π^p is a mapping on the p-simplices of K.

All possible patterns clearly constitute a <u>graded ring of polynomials</u> $\Pi(\Lambda V)$ on ΛV, with the same grading convention as before, viz.,

$$\Pi = \Pi^{-1} + \ldots + \Pi^{n} \quad (n \geqslant N)$$

in which the <u>ring product</u> is such that $\pi^p \in \Pi^p$ and $\pi^q \in \Pi^q$ result in

$$\pi^p \, . \, \pi^q \;\; \pi \;\; \Pi^{p+q} \quad \text{if } p + q \leqslant n$$

and

$$\pi^p \, . \, \pi^q = 0 \qquad \text{otherwise.}$$

We notice, however, that the product of two patterns π_1, π_2 on a fixed K is not necessarily a pattern on that K (for example, if $\pi_1 \equiv \pi_1{}^p$, $\pi_2 \equiv \pi_2{}^q$ and $p + q > N$).

<u>Example</u>: <u>The Board Game of Noughts and Crosses</u> (Figure 3)

The relation Γ between the nine squares $\{S_i\}$ and the eight lines $\{L_j\}$ is simply defined by $S_i \; \Gamma \; L_j$ if S_i lies on L_j.

Imagine the game is played between two opponents, White and Black, White to play first (placing the crosses). Denote the state of play by the "mode" [p, q]", indicating that White has placed p crosses and Black q noughts. Then Γ is a function of the mode, and so the structures $K_s(L; \Gamma)$ and K_L $(S; \Gamma^{-1})$ are also functions of the mode. We consider $K_L(S)$ in a game which White wins (by faulty play on Black's behalf) and represent the status of White's game (and of Black's) by a pattern polynomial on this $K_L(S)$. The algebraic variables in ΛV will be $s_1, s_2, \ldots, s_9$.

When White places a cross in square S_i then this square has been removed from Bäack's view of the board, and so the structure of the game (from Black's side) has altered. Precisely, if White occupies S_1 then, for Black, L_1 becomes the simplex $\langle s_2 s_3 \rangle$ instead of $\langle s_1 s_2 s_3 \rangle$. Also, placing a cross in s_1, which is the simplex $\langle L_1 L_4 L_8 \rangle$ in $K_s(L)$, can be represented by the pattern polynomial

$$\pi = L_1 + L_4 + L_8$$

on $K_L(S)$. White wins when, say, his play results in a polynomial π which contains the term $3L_j$, for some j. This is only possible if L_j is a 2-simplex $\langle s_\alpha\ s_\beta\ s_\gamma \rangle$ in the appropriate $K_L(S)$, in the relevant mode. In other words, White wins if

(a) π contains the term $3L$,

(b) π is graded so that $L_j \in \Pi^2$.

Conversely, Black frustrates White's play by altering the geometry (and so the algebraic structure) so that the proposed L_j is of lower dimension. The play might therefore proceed as follows.

Mode [0, 0], the structure for White (or Black) is $K_L(S)$ and represented algebraically by

$$\pi_o^W = s_1s_2s_3+s_4s_5s_6+s_7s_8s_9+s_1s_4s_7+s_2s_5s_8+s_3s_6s_9+s_3s_5s_7+s_1s_5s_9.$$

Mode [1, 0], after White has occupied square S_1: White's structure is given by $\pi_1^W = \pi_o^W$, plus the patter

$$\pi_W^2 = L_1 + L_4 + L_8 = s_1s_2s_3+s_1s_4s_7+s_1s_5s_9$$

Black's structure has altered so that it is π_o^W without the variable s_1, that is, by

$$\pi_1^B = s_2s_3+s_4s_5s_6+s_7s_8s_9+s_4s_7+s_2s_5s_8+s_3s_6s_9+s_3s_5s_7+s_5s_9$$

Thus, for Black, L_1, L_4, L_8, have become 1-simplices in $K_L(S)$. It is therefore impossible for Black to obtain a pattern containing $3L_j$, where $j = 1, 4, 8$.

Mode [1, 1], after Black has occupied square S_i: White's structure loses s_7 and becomes

$$\pi_2^W = a_1s_2s_3+s_4s_5'+s_8s_9 \ +s_1s_4+s_2s_5s_8+s_3s's_9+s_3s_5+s_1s_5s_9$$

and his pattern becomes

$$\pi_W^1 + \pi_W^2 = L_4 + (L_1 + L_8)$$

Black's structure remains the same with $\pi_2^B = \pi_1^B$, together with the pattern

$$\pi_B^1 + \pi_B^2 = L_4 + (L_3 + L_7)$$

Mode [2, 1], after White has occupied S_9: White's game is described by $\pi_3^W = \pi_2^W$ together with

$$\pi_W^1 + \pi_W^2 = (L_3 + L_4) + (L_1 + 2L_8 + L_6)$$

Black's game is given by

$$\pi_3^B = s_2s_3+s_4s_5s_6+s_7s_8+s_4s_7+s_2s_5s_8+s_3s_6+s_3s_5s_7+s_5$$

plus

$$\pi_B^1 + \pi_B^2 = (L_4 + L_3) + L_7$$

Mode [2, 2], after Black has occupied S_5 (because of the term $2L_8$) White's game is given by

$$\pi_4^W = s_1s_2s_3+s_4s_6+s_8s_9+s_1s_4+s_2s_8+s_3s_6s_9+s_3+s_1s_9$$

plus

$$\pi_W^1 + \pi_W^2 = (L_3 + L_4 + 2L_8) + L_1 + L_6))$$

Black's game is given by $\pi_4^B = \pi_3^B$, plus

$$\pi_B^0 + \pi_B^1 + \pi_B^2 = L_8 + (L_3 + L_4) + 2L_7$$

Mode [3, 2], after White has occupies S_3 (because of $2L_7$ in π_B^2). White's game is given by $\pi = {}_5^w\pi_4^w$ plus

$$\pi_w^0 + \pi_w^1 + \pi_w^2 = L_7 + (L_3 + L_4 + 2L_8) + (2L_1 + 2L_6)$$

Black's game is given by

$$\pi_5^w = s_2+s_4s_5s_6+s_7s_8+s_4s_7+s_2s_5s_8+s_6+s_5s_7+s_5$$

plus

$$\pi_B^0 + \pi_B^1 + \pi_B^2 = L_8 + (L_3 + L_4 + 2L_7) + (L_2 + L_5)$$

Mode [3, 3], after Black has occupied S_2 (because of $2L_1$ in π_w^2). White's game is given by π_6^w plus

$$\pi_w^0 + \pi_w^1 + \pi_w^2 = L_7 + (L_3 + L_4 + 2L_8 + 2L_1) + 2L_6$$

Black's game is given by

$$\pi_6^B = \pi_5^B \text{ plus } \pi_B^0 + \pi_B^1 + \pi_B^2 = (L_1 + L_8) + (L_3 + L_4 + 2L_7) + (2L_5 + L_2)$$

Mode [4, 3], White wins by occupying S_6.

White's pattern becomes

$$\pi_w^0 + \pi_w^1 + \pi_w^2 = L_7 + (L_2 + L_3 + L_4 + 2L_1 + 2L_8) + 3L_6$$

Black's play was faulty on Mode [1, 1]. The correct play, which leads to a draw, was to occupy square S_5. This is because, in $K_s(L)$, S_5 has maximum dimension, in fact $S_5 = \langle L_2\ L_5\ L_7\ L_8\rangle$ and occupation by Black reduces the dimensions of these four lines in White's structure. This gives Black the initiative since the move also gives Black a component in π_B^2, viz., $(L_2 + L_5 + L_7)$. White must then waste his moves in order to frustrate Black's plans; correct play leads to a draw.

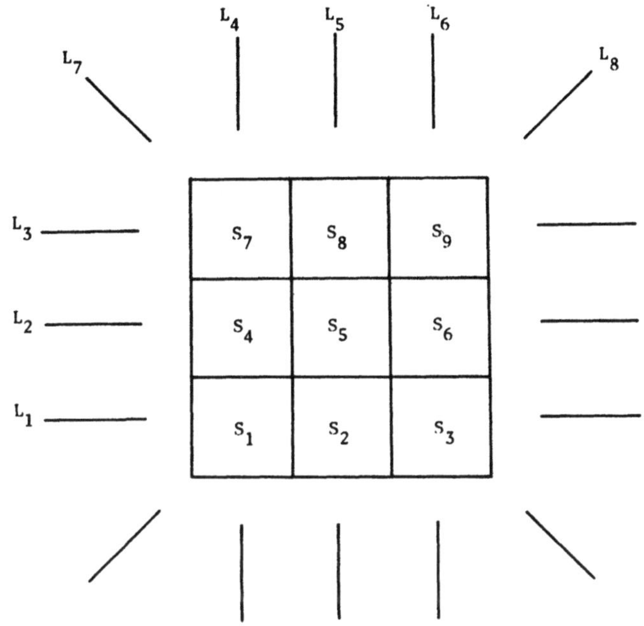

Figure 3

A2.3 The face operator f

The process of finding the faces of a simplex σ_p can be expressed in the algebraic representation by a suitable operator, f, defined as follows:

If $\sigma_p = \langle X_{\alpha_1}, X_{\alpha_2}, \ldots, X_{\alpha_{p+1}} \rangle$ where $\alpha_1 < \alpha_2 < \ldots < \alpha_{p+1}$ and $p > 0$

then $f\sigma_p = f(x_{\alpha_1}\, x_{\alpha_2} \ldots x_{\alpha_{p+1}})$

$$= \sum_{i=1}^{p+1} x_{\alpha_1}\, \hat{x}_{\alpha_2} \ldots x_{\alpha_i} \ldots x_{\alpha_{p+1}} \in \Pi^p$$

Thus, for example,

$$f(x_2x_4x_5) = x_4x_5 + x_2x_5 + x_2x_4, \ \varepsilon \ \Pi^2$$

[Notice that this is not to be confused with the boundary operator ∂ of conventional homology theory].

The face operator f now acts on a pattern polynomial π via linear extension to give a linear map f with the properties:

(i) $f(\alpha, \sigma_p) = \alpha(f\sigma_p)$, $\qquad \alpha \ \varepsilon \ J$

(ii) $f(\sigma_p + \sigma_q) = (f\sigma_p) + (f\sigma_q)$

taken together with

(iii) $f \ \sigma_o = \sigma_{-1} = 1 \ \varepsilon \ \Lambda^o V$

(iv) $f\sigma_{-1} = 0 \qquad \varepsilon \ \Lambda \ V$

We notice that if two simplices σ_p, σ_q share a vector then, in ΛV, $\sigma_p \cdot \sigma_q = 0$, since some x_i appears twice and $x_i^{\ 2} = 0$. Similarly, if they share two vertices (or a common σ_1) then not only is it true that

$$\sigma_p \cdot \sigma_q = 0$$

but also $\qquad \sigma_p \cdot (f\sigma_q) = (f\sigma_p) \cdot \sigma_q = 0$

although $\qquad \sigma_p \cdot (f^t \ \sigma_q) \neq 0$ if $t > 1$

Continuing this argument leads naturally to the following theorem, which expresses the pairwise connectivity between simplices:

<u>Theorem</u>: Two simplices σ_p, σ_q in K share a t-face if and only if their monomial representatives in ΛV satisfy,

either $(f^i \ \sigma_p) \cdot \sigma_q = 0$ for $i = 0, 1, \ldots, t$

$\neq 0$ otherwise.

or $\quad (\sigma_p) \cdot (f^i \sigma_q) = 0$ for $i = 0, 1, \ldots, t$

where $\ f^i = f(f^{i-1})$ and f^o = identity map.

We notice that the face operator f produces repetitions of the faces of any σ_p, when applied more than once. Although this is acceptable in the formulation of the above theorem nevertheless it can mean that unwanted coefficients are introduced by operations such as $f^t(\sigma_p)$, $t > 1$. We can avoid this situation by using, when appropriate, an associated "exponential" face operator $\hat{f}$, defined by

$$\hat{f}^t = \frac{1}{t!} f^t \quad t = 0, 1, \ldots, N$$

In $f^1(\sigma_p)$ each face, σ_{p-1}, occurs 1 time, whereas
in $f^2(\sigma_p)$ each face, σ_{p-2}, occurs 2.1 times,
in $f^3(\sigma_p)$ each face, σ_{p-3}, occurs 3.2.1 times, so that, when $t \leqslant p$,
in $f^t(\sigma_p)$ each face, σ_{p-t}, occurs $t!$ times.

It follows that, in ΛV,

$$\sum_{t=0}^{p} \hat{f}^t(\sigma_p) = \sum_{t=0}^{p} \frac{1}{t!} f^t(\sigma_p)$$

and this means that

$$\sum_{t=0}^{p} \hat{f}^t(\sigma_p) = \text{Sum of all the faces of } \sigma_p\text{, each occurring once only}$$

We can use this idea to produce an economical representation in ΛV of the complex $KY(X;\lambda)$. To begin with we notice that the elements Y_i are associated with the simplices of KY in the sense that every $\sigma_p \in \overset{o}{K}$ is a face of (at least) one Y_i. We can therefore select a subset of Y, say $\overset{o}{Y}$, such that

$$\text{if} \quad Y_i, Y_j \in \overset{o}{Y} \text{ then } Y_i \not\subset f\, Y_j$$

If we use the algebraic representation of these elements of $\overset{o}{Y}$ we have that if, $Y_i \equiv \sigma_p$, $Y_j \equiv \sigma_q$, then by the above theorem,

$$(f^p\, \sigma_p) \cdot \sigma_q \neq 0$$

in the case $p \leqslant q$.

These elements of $\overset{o}{Y}$ may therefore be regarded as characterising the q-components of $KY(X;\lambda)$, in the following way.

If KY consists of a single σ_N, $\overset{o}{Y}$ need contain only one element, viz., any one of the Y_i which are represented by that σ_N. This Y_i (say), together with all its faces (counted singly) constitutes the whole of KY. In this case we can write the representative polynomial as

$$\hat{\pi}_o = \sum_{t=0}^{N} \hat{f}^t(Y_1) = \sum_{t=0}^{N} \hat{f}^t(\sigma_N)$$

In general we require to make use of the triangular array given by the Q-analysis[1] relevant to $\overset{o}{Y}$. Each entry q_{ij} ($i \leq j$, $i = 1, \ldots, \text{card } \overset{o}{Y}$) in this array gives the dimension of the face shared by Y_i and Y_j. If we denote the shared q_{ij}-face by (Y_i, Y_j) we can see that the representative polynomial is, by the Principle of Inclusion-Exclusion[3],

$$\hat{\pi}_o = \sum_{i=1}^{m} \sum_{t=0}^{q_{ij}} \hat{f}^t(Y_i) - \sum_{i<j} \sum_{t=0}^{q_{ij}} \hat{f}^t(Y_i\ Y_j)$$

$$+ \sum_{i<j<k} \sum_{t=0}^{t_3} \hat{f}^t(Y_i\ Y_j\ Y_k) - \ldots + (-1)^m \sum_{t=0}^{t_m} \hat{f}^t(Y_1 \ldots Y_m)$$

with t_3 denoting the dimension of the face common to Y_i, Y_j, Y_k, etc., and this will be called the <u>canonical representation</u> in ΛV, defined by the characteristic <u>set</u> $\overset{o}{Y} = \{Y_1\ Y_2 \ldots Y_m\}$.

<u>A2.4 The coface operator Δ</u>

With the face operator f we can associate the dual coface operator, Δ, defined in the following.

Denote the image $\pi(\sigma_p) \in J$ by the inner product notation

$$(\sigma_p, \pi)$$

and let Δ be a map

$$\Delta : \Pi^t \to \Pi^{t+1}$$

such that

$$(f\sigma_p, \pi) = (\sigma_p, \Delta\pi)$$

If we denote $f \mid K^p$ by f_p then it is appropriate to write

$$\Delta = \{\Delta_p\}$$

where $$\Delta_p : \Pi^p \to \Pi^{p+1}$$

and

$$(f_p\sigma_p, \pi^{p-1}) = (\sigma_p, \Delta_{p-1}\, \pi^{p-1})$$

If $\pi = \pi^p$ is a graded pattern, associated exclusively with the p-dimensional simplices of K, then

$$\Delta\pi \equiv \Delta\pi^p \in \Pi^{p+1}$$

is a (p+1)-pattern, associated with those (p+1)-simplices whose p-faces are associated with π^p.

By requiring that linearity should be preserved under Δ we can naturally extend it to a homomorphism (also denoted by Δ),

$$\Delta : \Pi \to \Pi$$

on the ring of pattern polynomials on K. In this context we need to notice that, in ΛV,

$$\Delta(\pi_1 \cdot \pi_2) = (\Delta\pi_1) \cdot \pi_2 + \pi_1 \cdot (\Delta\pi_2)$$

$$\Delta^2(\pi_1 \cdot \pi_2) = (\Delta^2\pi_1) \cdot \pi_2 + 2(\Delta\pi_1) \cdot (\Delta\pi_2) + \pi_1 \cdot (\Delta^2\pi_2)$$

which suggests that if we write

$$\Delta = \Delta_1 + \Delta_2$$

then

$$\Delta^k = (\Delta_1 + \Delta_2)^k$$

in which we suppose

$$\Delta_1 = \Delta|\pi_1 \text{ and } \Delta_2 = \Delta|\pi_2$$

It also follows that we can interpret Δ^r, $r > 1$, by way of the definition of f^r. Thus

$$(f^r\sigma_{p+r}, \pi^p) = (f^{r-1}\sigma_{p+r}, \Delta\pi^p) = \text{etc.}$$

$$= (f^{r-s}\sigma_{p+r}, \Delta^s\pi^p)$$

$$= (f_{p+r}, \Delta^r\pi^p)$$

Thus, $\Delta^r\pi^p$ is a map defined on (p+r)-dimensional simplices. The mutual actions of f and Δ on the graded ring Π, are illustrated by the scheme:

$$\Pi^o \oplus \Pi^1 \underset{\underset{\Delta}{\rightarrow}}{\overset{\overset{f}{\leftarrow}}{\oplus}} \Pi^2 \ldots \oplus \Pi^p \underset{\rightarrow}{\overset{\leftarrow}{\oplus}} \Pi^{p+1} \oplus \ldots$$

If we move to the exponential face operator $\hat{f}$ we obtain a dual operator $\hat{\Delta}$, via $(\hat{f}\sigma_p,\pi) = (\sigma_p,\hat{\Delta}\pi)$. Since

$$(\hat{f}^t\sigma_p, \pi) = \frac{1}{t!}\,(f^t\sigma_p, \pi) = \frac{1}{t!}\,(\sigma_p, \Delta^t\pi)$$

it follows that

$$\hat{\Delta}^t = \frac{1}{t!}\,\Delta^t,\ t \geqslant 0$$

<u>Example</u>: Given a pattern polynomial

$$\pi = 2 + 4x_1x_2 + 7x_3x_4x_5x_6$$

then

$$f(\pi) = 0 + 4x_1 + 4x_2 + 7x_4x_5x_6 + 7x_3x_5x_6 + 7x_3x_4x_6 + 7x_3x_4x_5$$

and

$$\Delta\pi = n_o^i\,\sigma_o^i + n_2^j\,\sigma_2^j + n_4^k\,\sigma_4^k$$

where repeated subscripts imply summation, and where i, j, k take all the relevant values such that, in K,

$$1 \leq \sigma_o^i,\ x_1x_2 \leq \sigma_2^j;\ x_3x_4x_5x_6 \leq \sigma_4^k$$

These sets will be denoted by $\{\sigma_o^i\} = f_{-1}{}^{(1)}$, $\{\sigma_2^j\} = f_{-1}(x_1x_2)$, and $\{\sigma_4^k\} = f_{-1}(x_3x_4x_5x_6)$. The numbers n_4^k must be such

$$\sum_n n_4^k = 7 = \pi \mid x_3x_4x_5x_6$$

We notice that, whereas,

$$f^2\pi = 8 + 14\ x_4x_5 + 14\ x_3x_4 + 14\ x_5x_6 + 14\ x_3x_5 + 14\ x_4x_6 + 14\ x_3x_6$$

we have

$$\hat{f}^2\pi = 4 + 7\ x_4x_5 + 7\ x_3x_4 + \ldots + 7\ x_3x_6$$

Similarly $\Delta^2\pi$ takes the same value on (say) $x_1x_2x_5x_6$ as π takes an $f^2(x_1x_2x_5x_6)$. This value is $8 = 2.4$; by comparison $\hat{\Delta}^2\pi$ takes the value 4 on $x_1x_2x_5x_6$ since π takes that value on $\hat{f}^2(x_1x_2x_5x_6)$.

We can see from this example that the specific role of Δ depends on the structure of K. In the special case that K is a simplex σ_{n-1} $(N = n-1)$, together with all its faces, then $f_{-1}(\sigma_p)$ in K is easily constructed, as follows:

If σ_o^i is a specific 0-simplex it has a representation x_i in ΛV, and $f^{-1}x_i$ consists of all the 1-simplices (edges) incident on x_i. There are (n-1) of these, for any given x_o

$$\text{card}\ \{f_{-1}\ x_i\} = (n-1)$$

If σ_1 has the representation x_ix_j in ΛV, fixed i, j, then $f_{-1}(x_ix_j)$ consists of all the 2-simplices (triangles) possessing x_ix_j as a face. There are (n-2) of these,

$$\text{card}\ \{f_{-1}(x_ix_j)\} = (n-2)$$

and so on.

If π^1 is a pattern on the 1-simplices it takes a value on each of the $\frac{1}{2}n(n-1)$ products $x_i x_j$. Subsequently, $\Delta\pi^1$ is defined on each of the $\frac{1}{2}n(n-1)(n-2)$ products $x_i x_j x_k$. Suppose, in particular, that π^1 takes the values $\alpha_r \in J$, $r = 1, 2, \ldots, \binom{n}{2}$, on the possible σ_1^i. Then each value α_r must make a contribution to $(n-2)$ σ_2's. If, in general, there are N_2^i σ_2's for which a particular $\sigma_1^i < \sigma_2$ then we are concerned with the manner in which the values α_i may be distributed over these σ_2's. Precisely, this means that there is a partition of α_i over these σ_2's; that is to say, numbers α_i^j $(j = 1, 2, \ldots, N_2^i)$ exist such that

$$\alpha_i = \sum_j \alpha_i^j \quad \text{and} \quad (\alpha_2^j, \Delta\pi^1) = \alpha_i^j$$

As a special case we can allow each α_i^j to be the same, for all j, and write,

$$\alpha_i^j = \frac{1}{N_2^i}\alpha_i = K$$

so that $\Delta\pi^1$ takes a constant value on each $\sigma_2 \in f_{-1}\,\sigma_1^i$. Precisely,

$$(\alpha_2, \Delta\pi^1) = k$$

In the case of a single simplex $\sigma_{n-1} = K$ we obtain

$$N_2 \,.\, k = \sum_{i=1}^{N_2} \alpha_i$$

where $N_1 = \binom{n}{2}$ and $N_2 = \binom{n}{3}$.

In this case we obtain the general result, viz.,

Given that π^p takes values α_r, $r = 1, 2, \ldots, N_p$, where $N_p = \binom{n}{p+1}$, then $\Delta\pi^p$ takes the value k on each σ_{p+1} where

$$N_{p+1} \,.\, k = \sum_{r=1}^{N_p} \alpha_r$$

where $N_{p+1} = \binom{n}{p+2}$, for $0 \leq p \leq n - 2$.

Thus α_r, on some σ_r, gives rise to a value k on every σ_{r+1} for which $\sigma_r \leq \sigma_{r+1}$, provided $r \leq n - 2$.

In the special case that $p = n - 1$, π^p is defined on the one σ_{n-1} in K. Now there does not exist any $\sigma_n \in K$ to "carry" the value of $\Delta\pi^p$. Hence $\Delta\pi^p$ is not defined in K.

We see how in this sense, the complex K offers an obstruction to the formation of $\Delta\pi^{n-1}$. This obstruction is inherent in the structure of K, in its abstract geometry.

If each $\alpha_r = 1$, at each p-level, the pattern polynomial is π_o, defining K itself. In this case

$$k = N_p/N_{p+1}, \; p = 0, 1, \ldots, (n-2)$$

with an obstruction at $p = (n-1)$ - which we may regard as arising algebraically out of the condition that N_{p+1} (i.e. N_{n+1} at $p = n$) vanishes. Thus, speaking loosely, the obstruction in K can only be overcome by $\Delta\pi^n$ taking an infinite value.

A2.5 The role of the obstruction vector $\hat{Q}$

A Q-analysis of the complex K gives us a first structure vector

$$\underset{\sim}{Q} = \{Q_N, Q_{N-1}, \ldots, Q_1, Q_0\}$$

where Q_r is the number of components at the r-level of connectedness.

Suppose σ_r is an r-simplex in one of the Q_r components but that it is not a face of any σ_{r+1} simplex; thus there is no (r+1)-simplex σ_{r+1} such that

$$\sigma_r < \sigma_{r+1}$$

Then it is precisely this circumstance that results in Q_r taking a value greater than unity.

It follows that if $(\sigma_r, \pi^r) \in J$ then $\Delta\pi^r$ is not defined with respect to this σ_r. Thus, the value of π_r on σ_r cannot be redistributed over those σ^i_{r+1} for which $\sigma_r \leq \sigma^i_{r+1}$ since they do not exist. Hence this kind of σ_r acts as an obstruction to allowed changes $\Delta\pi^r$. The obstruction is inherent in the geometry of K.

If $Q_o = 1$ and $Q_N = 1$, K is an N-simplex σ_N then the only obstruction occurs at $r = N$. There is zero obstruction at all levels r such that

$$0 \leq r < N$$

that is to say, the geometry of K freely allows $\Delta\pi^r$, given π^r.

If $Q_o = 1$, but K is not a single N-simplex then there exists a non-trivial filtration of K such that

$$K = \bigcup_p K^p \quad p = 0, 1, \ldots, N$$

and

$$K^N \subset K^{N-1} \subset \ldots \subset K^p \subset K^{p-1} \subset \ldots \subset K^1 \subset K^o = K$$

We can now form a graded complex K by

$$K = \bigoplus_p K_p$$

and $K_p = K^p/K^{p+1}$, consisting of classes - each one of which has a representative p-simplex which is not a face of any (p+1)-simplex. It follows that if $Q_r > 1$, $0 < r < N$, there are exactly $(Q_r - 1)$ r-simplices which act as obstructions to Δ.

Thus we see that the obstruction vector

$$\hat{\underline{Q}} = \{Q_N, Q_{N-1} - 1, Q_{N-2} - 1, \ldots, Q_1 - 1, 0\}$$

is a measure of the obstruction to Δ, at any level r such that $0 < r < N$. In a previous paper[1] the obstruction vector was written so as to include the component $(Q_N - 1)$ but whereas this is harmless for $r < N$ we can see that, in general, such a component disguises the existence of obstruction at the N-level.

Finally, we see that when $Q_o > 1$ then K consists of Q_o disjoint complexes $K^{(i)}$ $i = 1, \ldots, Q_o$, and the above argument applies to each $K^{(i)}$ separately.

A2.6 Incremental changes $\delta\pi$ in a pattern π

Each pattern π on a complex K, regarded as a polynomial in the exterior algebra $\Lambda(X_n)$, may be associated with a pattern $\Delta\mu$ where Δ is the coface operator. If $\sigma_p \in K$ then we need only write

$$(\sigma_p, \pi) = \sum_i (f_{-1}\sigma_p, \Delta\mu)$$

where $f_{-1}\sigma_p = \{\sigma^i_{p+1}; \sigma_p \leq \sigma^i_{p+1}\}$.

This means that $\Delta\mu$ is the pattern, associated to the $f_{-1}\sigma_p$, the (p+1)-level, which corresponds to the (given) pattern π at the p-level.

Now an increment $\delta\pi$, in a pattern π, may thereby be associated to such a $\Delta\mu$, via

$$(\sigma_p, \delta\pi) = (f_{-1}\,\sigma_p, \Delta\pi) \text{ for all p.}$$

In the applications of the idea of patterns[2] we interpret an incremental change $\delta\pi$ as a <u>force</u> in the structure; if $\delta\pi \equiv (\delta\pi)^t$ then we speak of a <u>t-force</u> in the complex K. This t-force is therefore associated to a pattern $(\Delta\mu)$ defined at the (t+1)-level. It is in this sense that we interpret the incremental $\delta\pi$ in the abstract geometry of K.

A motivation for this association arises via a consideration of the notion of δv, where v is a velocity (a 1-pattern) defined on a σ_1 (say) $\langle P_1\ P_2\rangle$. Then δv is a measure of a "force" throughout an interval of time δt, and this δv is associated to a triangle (a σ_2) of which $\langle P_1, P_2\rangle$ is a face. Thus we are saying that the physicist's "force" δv is naturally tied to a 2-pattern defined on (say) $\langle P_1\ P_2\ P_3\rangle$ via

$$(\langle P_2\ P_3\rangle, \delta v) = (\langle P_1\ P_2\ P_3\rangle, \Delta\mu)$$
$$= (\langle P_1\ P_2\ P_3\rangle, \text{force}),$$

and this is illustrated in Figure 4.

The <u>intensity</u> of a t-force, on a particular σ_t, is defined as

$$(\sigma_t, (\delta\pi)^t) \div (\sigma_t, \pi^t), \text{ when that exists}$$

and this is associated to the ratio

$$\sum_i (f_{-1}\,\sigma_t, \Delta\mu) \div (\sigma_t, \pi^t)$$

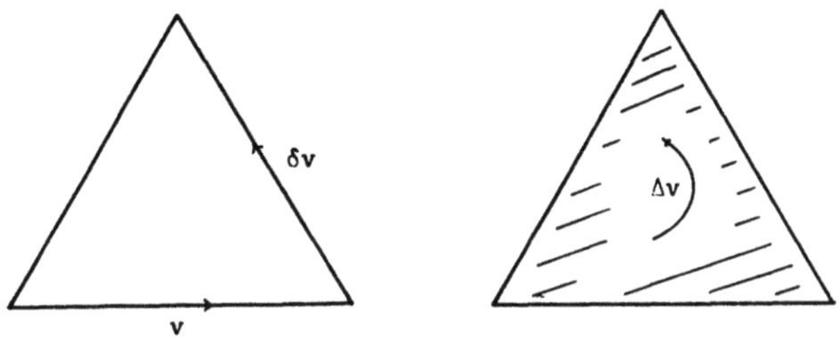

Figure 4 Δ associated with δ

A2.7 Extending the exterior algebra ΛV to Ex(V)

Given the vertex set X, card X = n, of an augmented oriented complex K then we have already formed an algebraic representation of K in the exterior algebra ΛX - patterns being polynomials $\pi \in \Lambda X$. Now, for a given vertex X_i consider the anti-vertex $\bar{X}_i$, which corresponds to not-X_i. Thus, if X_i is an urban-structure-type vertex representing "residence", $\bar{X}_i$ would represent the presence of "not-residence" (for example, "factory" or "shop").

Using this set $\bar{X}$ of anti-vertices we can construct a complementary complex $\bar{K}$ which is to be associated with the original complex K. Since K is defined by the relation $\lambda \subset Y \times X$ we can suppose that the new complex $\bar{K}$ is specifically defined by an associated relation $\bar{\lambda} \subset Y \times \bar{X}$. Thus $\bar{\lambda}$ will be such that $Y_i \bar{\lambda} \bar{X}_j$ whenever $Y_i \lambda X_j$, and conversely. Thus the incidence matrix $\bar{\Lambda}$, of $\bar{\lambda}$, will be the complement of the incidence matrix Λ, of λ. By extending the set X to the set $X \cup \bar{X}$ we obtain an incidence matrix $\tilde{\Lambda}$, obtained by simply placing Λ, $\bar{\Lambda}$ side by side. Thus we obtain a well-defined relation $\tilde{\lambda} \subset Y \times (X \cup \bar{X})$, induced from the initial $\lambda \subset Y \times X$. In this case, when λ induces $\tilde{\lambda}$, each simplex $Y_i \in KY(X_i\lambda)$ becomes an n-simplex $\tilde{\sigma}_n \in KY(X \cup \bar{X}); \tilde{\lambda})$, because if Y_i is λ-related to a (p+1)-subset of X it is $\bar{\lambda}$-related to the remaining (n-p-1)-subset of $\bar{X}$. It follows that, when λ induces $\tilde{\lambda}$, dim Y_i = (n-1), a constant, for all i = 1, ..., card Y.

But then in addition to this case we now have the possibility of a new combined complex $\tilde{K}$ defined by <u>any relation</u> $\tilde{\lambda} \subset Y \times (X \cup \bar{X})$, between the sets Y and $X \cup \bar{X}$. Then if we consider the possibility of the simplex $\langle x_i, \bar{x}_i \rangle$ in $\tilde{K}Y(X \cup \bar{X}; \tilde{\lambda})$ we would require some $Y_i \in Y$ such that $Y_j \tilde{\lambda} X_i$ <u>and</u> $Y_j \tilde{\lambda} \bar{X}_i$. This means that $Y_j = \sigma_{-1}$, the empty set. In an algebraic representation of $\tilde{K}Y(X \cup \bar{X}; \lambda)$ the composition of $\bar{x}_i$ with x_i must therefore be equivalent to the unity, 1. To ensure this situation we therefore introduce an extended algebra to accommodate this extended (or <u>complemented</u>) complex as follows.

<u>Definition 1</u> Given an n-set X and an associated exterior algebra ΛX, over a ring J, we define the <u>extended exterior algebra</u>, Ex $(X \cup \bar{X})$, by introducing a basis element $\bar{x}_i$ for each i = 1, ..., n, with the properties:

(i) $\bar{x}_i \cdot x_i = x_i \cdot \bar{x}_i = 1$, i = 1, 2, ..., n

(ii) $\bar{x}_i \cdot x_j = -x_j \cdot \bar{x}_i$, $i \neq j$

(iii) $\bar{x}_i \cdot \bar{x}_j = -\bar{x}_j \cdot \bar{x}_i$; $x_i \cdot x_j = -x_j \cdot x_i$; for all i, j.

where the " . " is the product of Ex.

It follows that Ex $(X \cup \bar{X})$ contains the sub-algebras ΛX and $\Lambda\bar{X}$, but it is not the direct sum of these. Furthermore, every oriented, augmented, complemented complex $\tilde{K}$ (abbreviated to "OAC-complex") possesses an algebraic representation in Ex $(\tilde{K})$. This representation is via a polynomial,

$$\pi^o = \sum_{p=0}^{m} \varepsilon_i \tilde{\sigma}_p^{\ i}$$

where $\varepsilon_i = 0, 1$; $m \leqslant n - 1$ and where $\tilde{\sigma}_p^{\ i} \in \tilde{K}$; each $\tilde{\sigma}_p^{\ i}$ being represented by a monomial of the form

$$x_{\alpha_1} x_{\alpha_2} \ldots x_{\alpha_s} \bar{x}_{\beta_1} \bar{x}_{\beta_2} \ldots \bar{x}_{\beta_{p-s+1}}$$

in Ex $(\tilde{K})$. When $s = p + 1$ this $\tilde{\sigma}_p^{\ i} \in K$, with a representation in $\Lambda(K)$, and when $x = 0$ this $\sigma_p^{\ i} \in \bar{K}$, with a representation in $\Lambda(\bar{K})$. But since there might generally be some $\tilde{\sigma}_p^{\ i}$ for which $0 < s < p + 1$ we have

$$\text{Ex } (\tilde{K}) \neq \Lambda(K) + \Lambda(\bar{K})$$

Such a simplex, with $0 < x < p + 1$, we might refer to as a <u>bridge simplex</u>, and say it is of <u>weight</u> s <u>in</u> $\Lambda(X)$.

Given a pattern polynomial $\pi \equiv \pi^o$, defined on each vertex in K, we can construct a pattern $\Delta\pi^o \in \Pi^1$- which is compatible with the geometrical structure of K - by using

$$(\hat{f}\ \sigma_1, \pi^o) = (\sigma_1, \hat{\Delta}\pi^o)$$

for each $\sigma_1 \in K$. Containing in this way we can construct the pattern

$$\pi = \pi^o + \hat{\Delta}\pi^o + \hat{\Delta}^2\pi^o + \ldots + \hat{\Delta}^N\pi^o$$

where N = dim K. The structure of K (because it contains an N-simplex) therefore permits patterns of the form

$$\pi = \{1 + \frac{\Delta}{1!} + \frac{\Delta^2}{2!} + \ldots + \frac{\Delta^N}{N!}\}\ \pi^o$$

writing e_k^x for the <u>truncated exponential</u> series (with last term $\frac{x^k}{k!}$) we can express this as

$$\pi = e_N^{\Delta}\ (\pi^o) \qquad \ldots \qquad (1)$$

The <u>ideally free structure</u>, with dim K = ∞, would be characterised by the existence of patterns of the form

$$\pi = e^{\Delta}(\pi^{o}) \qquad \dots \qquad (2)$$

for arbitrary π^{o}. Furthermore, for arbitrary π^{t} the <u>pattern generator</u> (1) will become

$$\pi = e^{\Delta}_{N-t}(\pi^{t}) \qquad \dots \qquad (1a)$$

whilst (2) becomes

$$\pi = e^{\Delta}(\pi^{t}) \qquad \dots \qquad (2a)$$

Equation (2a) implies $\pi^{t} = e^{-\Delta}(\pi)$

but this is not applicable to (1a).

If we now assume that K is characterised by the set $\overset{o}{Y}$ (v. section A2.3) with card Y = s, then there is a finite set of simplices $\{Y_i\}$ with dimensions q_{ii} (max q_{ii} = N) whose intersections two, three, ... at a time have dimensions q_{ij}, q_{ijk}, etc.. The canonical representation of K now induces an overall pattern generator of the form

$$\Theta(K) = \sum_{i=1} e^{\Delta}_{q_{ii}} - \sum_{i<j} e^{\Delta}_{q_{ij}} + \sum_{i<j<k} e^{\Delta}_{q_{ijk}} - \text{etc..} \qquad \dots \qquad (3)$$

with the proviso that when q_{ij} = -1 the corresponding j-term is eliminated from the sum, and so on for triples etc.. This operator Θ(K) characterises the geometrical structure of the complex K. Acting on polynomials $\pi \in \Pi$ it generates those polynomials which are compatible with that geometry.

An alternative form for Θ(K) can be obtained by noticing that, at any t-level, the coefficient of $\frac{\Delta t}{t!}$ in (3) must equal the number of separate components at that level. This number is Q_t, where

$$\underset{\sim}{Q} = \{Q_N, Q_{N-1}, \dots, Q_1, Q_0\}$$

is the structure vector of section A1.6. This situation is illustrated in the "sky-scraper" diagram of the accompanying Figure 5.

In this figure the separate "sky-scrapers" are joined at a level corresponding to $q = q_{ij}$, although it cannot show the total of all connections between subsets of the $\overset{o}{Y}_i$.

The expression for $\Theta(K)$ is therefore equivalent to

$$\Theta(K) = \sum_{t=0}^{N} Q_t \cdot \frac{\Delta t}{t!} \qquad \ldots \quad (3a)$$

or

$$\Theta(K) = \sum_{t=0}^{N} Q_t \cdot \hat{\Delta}^t \qquad \ldots \quad (3b)$$

This can also be regarded as the formal scalar product (in Q-space) of $\underline{Q}$ with, for example, the face vector operator $\underline{\hat{\Delta}}$ where

$$\underline{\hat{\Delta}} = \{\hat{\Delta}^N, \hat{\Delta}^{N-1}, \ldots, \hat{\Delta}^o\}$$

so that we have the symbolic

$$\Theta(K) = \underline{Q} \cdot \underline{\hat{\Delta}} \qquad \ldots \quad (4)$$

The role of the obstruction vector $\underline{\hat{Q}}$ now emerges by comparing (4) with that due to a single simplex σ_N, viz.,

$$\Theta(\sigma_N) = \underline{1} \cdot \underline{\hat{\Delta}}$$

and then, with the understanding that $Q_o = 1$,

$$\Theta(K) - \Theta(\sigma_N) = \underline{\hat{Q}} \cdot \underline{\hat{\Delta}} \qquad \ldots \quad (5)$$

The number $Q_t = Q_t - 1$ measures the number of gaps between the sky-scrapers (the $\overset{o}{Y}_i$) at the t-level; patterns $\Delta^{t-1}\pi$ cannot be extended to $\Delta^t\pi$ <u>across</u> the gaps.

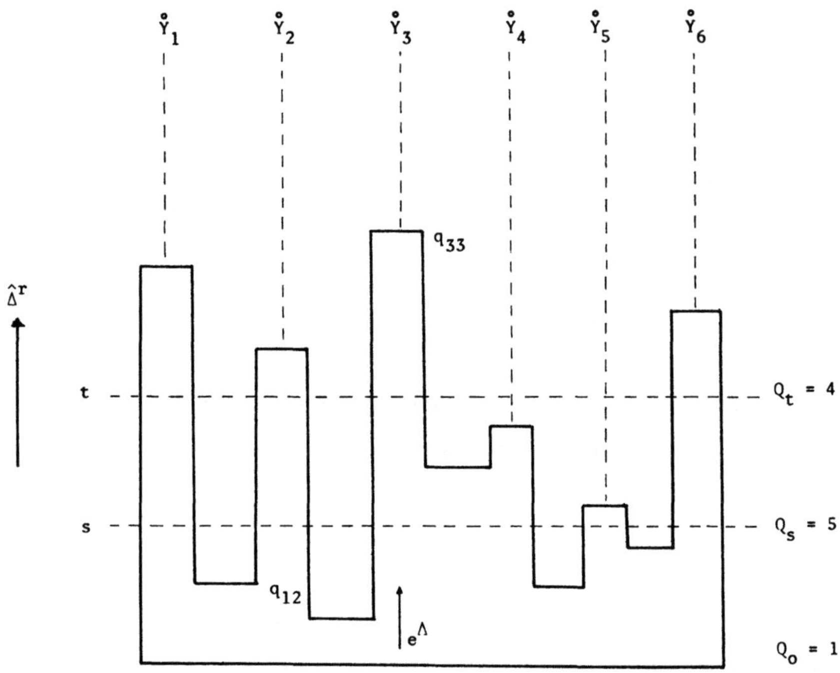

Figure 5. Skyscraper model for $\Theta(K)$

Example: Car Parking patterns in a town centre

In the time interval 0830-1100 hrs., on a typical working day, the car parking in a certain area of a town constituted the following polynomial pattern,

$$\pi_1 = 4x_1 + 17x_2 + 46x_3 + 7x_4 + x_5 + x_6 + 3x_7 + 3x_8 + 4x_1x_2 + x_1x_6$$
$$+ 2x_2x_3 + x_2x_6 + 2x_1x_2x_4 + 3x_1x_2x_6 + x_3x_4x_5x_6 + x_1x_2x_4x_5x_6$$

where the structure variables $\{x_i; i = 1, \ldots, 8\}$ refer to a set of vertices describing available activities which the motorist wished to use. Thus π_1 is graded into

$$\pi_1 = \pi^0 \oplus \pi^1 \oplus \pi^2 \oplus \pi^3 \oplus \pi^4$$

where, for example, $\pi^0 = 4x_1 + 17x_2 + 46x_3 + 7x_4 + x_5 + x_6 + 3x_7 + 3x_8$.

The meaning of the term $3x_1x_2x_6$ in π_1 is that 3 motorists parked their cars in this area in order to make use of the amenities corresponding to the vertices X_1, X_2, X_6.

For the time interval 1200 - 1430, on the same day, the corresponding pattern polynomial was

$$\pi_2 = 2x_1 + 3x_2 + 11x_3 + 2x_4 + x_5 + 2x_6 + 4x_8 + 2x_1x_2 + x_1x_6 + x_2x_3 + x_1x_2x_3x_4x_6$$

being graded into

$$\pi^2 = \pi^0 \oplus \pi^1 \oplus \pi^4$$

Let us suppose that this area of study contains all eight features corresponding to the variables $x_1, \ldots, x_8$. Then the area possesses the structure of a 7-simplex, viz., $x_1x_2x_3x_4x_5x_6x_7x_8$, together with all its faces.

The structure vector is therefore

$$\underset{\sim}{Q} = \{\overset{7}{1} \quad 1 \quad 1 \quad 1 \quad 1 \quad 1 \quad 1 \quad \overset{0}{1}\}$$

and the pattern generator $\Theta(\sigma_7)$ is, by equation (4) above,

$$\Theta = \underset{\sim}{Q} \cdot \underline{\hat{\Delta}} = \{\hat{\Delta}^7, \hat{\Delta}^6, \ldots, \hat{\Delta}^0\}$$

This generates an "allowed" pattern from every π^0, that is to say, every π^0 is geometrically compatible with $\Theta\pi$. In the same way it is clear that, for example, every π^4 is compatible with

$$\{\hat{\Delta}^3, \hat{\Delta}^2, \hat{\Delta}^1, \hat{\Delta}^0\} \; \pi^4$$

which we might write as $\Theta_{(4)} \; \pi^4$

By forming the pattern

$$\Theta \; \pi_1^0 + \Theta_{(1)}\pi_1^1 + \Theta_{(2)}\pi_1^2 + \Theta_{(3)}\pi_1^3 + \Theta_{(4)}\pi_1^4 \qquad \ldots \quad (6)$$

from the observed π_1 we obtained a total pattern compatible with the abstract geometry. The significance of this for the car parking "problem" of the area (σ_7) lies in its relation to the physical constraints imposed by the local traffic engineer. In effect, he lays down a certain criterion of parking tolerances in the area. Let this be a crude total number N of car parking spaces. Then the pattern π_1 indicates that

$$N \geqslant M \qquad \ldots \quad (7)$$

where M = sum of all the polynomial coefficients in (6).

Thus, the physical control of traffic in this area (σ_7) must provide for a condition like (7) if the physical development is to be matched with the implications of the abstract structure given by (6). The same considerations apply, of course, to π_2 etc. as well as to all the other areas of the town.

Before leaving this example we can illustrate the notion of "structural force" which is inherent in the change of pattern $\pi_1 \to \pi_2$. Defining

$$\delta\pi = \pi_2 - \pi_1 = -2x_1 - 14x_2 - 35x_3 - 5x_4 + x_6 - 3x_7 + x_8$$

$$- 2x_1x_2 + x_1x_3 - 2x_2x_3 - x_2x_6 - 2x_1x_2x_4 - 3x_1x_2x_6$$

$$+ x_1x_2x_3x_4x_6 - x_1x_2x_4x_5x_6$$

we obtain 0-forces of repulsion on the vertices X_1, X_2, X_3, X_4, X_7 and 0-forces of attraction on the vertices X_6, X_8. Similarly we obtain 1-forces of repulsion on $\langle X_1X_2\rangle$, $\langle X_2X_3\rangle$, $\langle X_2X_6\rangle$ and a 1-force of attraction on $\langle X_1X_3\rangle$. The intensities of, for example, the 1-forces of repulsion, are

$$^2/_4 \text{ on } \langle X_1X_2\rangle, \quad 1 \text{ on } \langle X_2X_3\rangle, \quad 1 \text{ on } \langle X_2X_6\rangle.$$

Since the geometry (of σ_7) allows the existence of the pattern $\Delta\mu$ corresponding to $\delta\pi$ these forces are internal and compatible with the structure. Thus the car-parking-motorists who take part in this change $\delta\pi$ do so unwittingly, without experiencing any sense of structural stress. The same would not be true, if for example, the change in $\pi_1{}^4$ were caused by the physical removal of the vertex X_5, for then the structural change would "cause" the motorist to move on (between the two time intervals) in order to find X_5 elsewhere. This causative event would be experienced by the motorist as a structural (or social) force.

A3.0 A concept of nearness

We shall introduce a notion of nearness between simplices and then go on to use this to analyse the proximity of chains of connection.

Definition: If two simplices σ_r, σ_s share a q-face, that is to say

$\sigma_r \cap \sigma_s = \sigma_q$, we shall say that they are q-near.

It follows that if σ_r, σ_s are q-near they must also be t-near, for $t = 0, 1, \ldots, q$.

The idea of a chain of q-connectivity between σ_r and σ_s may now be equally well expressed in terms of q-nearness. In fact, if σ_r is q-connected to σ_s then there exists a finite sequence of simplices $\{\sigma_{\alpha_i}; i = 1, \ldots, h\}$ such that (i) σ_r is q-near σ_{α_1}

(ii) σ_s is q-near σ_{α_h}

(iii) σ_{α_i} is q-near $\sigma_{\alpha_{i+1}}$, $i = 1, \ldots, (h-1)$

Let us use the notation $[\sigma_r, \sigma_s]$ for a chain of connection between σ_r and σ_s. When it is important to indicate the order, q, of this chain, we can write $[\sigma_r, \sigma_s]$. Given two chains of connection $[\sigma_r, \sigma_s]$ and $[\sigma_h, \sigma_k]$ in a complex K, we can use the above concept of nearness to introduce the idea of two chains being adjacent.

Definition: The chains of connection $[\sigma_r, \sigma_s]$ and $[\sigma_h, \sigma_k]$ are said to be q-adjacent if

(i) σ_r is q-near σ_h

(ii) σ_s is q-near σ_k

(iii) the sequence $\{\sigma_t\}$, $\{\sigma_m\}$ which define the chains $[\sigma_r, \sigma_s]$, $[\sigma_h, \sigma_k]$, are separately ordered (denote the ordering by $\lessdot$)

(iv) for every $\sigma_t \in [\sigma_r, \sigma_s]$ there exists a $\sigma_m \in [\sigma_h, \sigma_k]$ such that σ_t is q-near to σ_m, and if $\sigma_{t_1} \lessdot \sigma_{t_2}$ then $\sigma_{m_1} \lessdot \sigma_{m_2}$.

(v) for every $\sigma_m \in [\sigma_h, \sigma_k]$ there exists a $\sigma_t \in [\sigma_r, \sigma_s]$ such that σ_m is q-near σ_t, and if $\sigma_{m_1} \lessdot \sigma_{m_2}$ then $\sigma_{t_1} \lessdot \sigma_{t_2}$.

It follows that, for example $[\sigma_r, \sigma_s]_q$ is q-adjacent to itself, and also that the relation of "being q-adjacent to" is symmetric - but not transitive.

A3.1 A form of discrete pseudo-homotopy

We can express a chain of connection $[\sigma_r, \sigma_s]$ by way of a mapping, c, in the following way.

Let J denote the additive group of integers and suppose that the chain $[\sigma_r, \sigma_s]$ contains n simplices σ_{α_i} for which $\sigma_{\alpha_o} = \sigma_r$ and $\sigma_{\alpha_{n-1}} = \sigma_s$. Regard the complex K as augmented by adjoining the (-1)-simplex σ_{-1}. Then $[\sigma_r, \sigma_s]$ can be identified with the mapping

$$c : J \to K^+ = K \cup \sigma_{-1}$$

where $\quad c(i) = \sigma_{\alpha_i}, \; i = 0, 1, \ldots (n-1)$

and $\quad c(i) = \sigma_{-1}$, otherwise.

This numbering of the simplex members of $[\sigma_r, \sigma_s]$ induces a natural <u>ordering</u> of them, that of the natural integers. The <u>reverse</u> ordering can then be induced by the mapping $\tau : J \to K^+$ where

$$\tau(j) = c(n - j - 1)$$

A chain of connection (or mapping) c is thus characterised by the initial and final simplices, σ_r and σ_s, and by the integer n (so that (n-1) may be regarded as the "length" of the chain - not counting σ_r in this). When these need to be specified we can refer to the chain of connection by $c\{\sigma_r, \sigma_s, n\}$.

Now two chains $c_1\{\sigma_r, \sigma_s, n_1\}$ and $c_2\{\sigma_h, \sigma_k, n_2\}$ might be related by a sequence of chains which are pairwise adjacent - this being an obvious extension of the definition of a chain of connection itself. This idea is a combinatorial analogue of the notion of homotopy[5] - in topology, and we shall refer to it as pseudo-homotopy - giving it the name of shomotopy and defining it via the following.

A (discrete) q-shomotopy between $c_1\{\sigma_r, \sigma_s, n_1\}$ and $c_2\{\sigma_h, \sigma_k, n_2\}$ is a mapping

$$Sh : J \times J \rightarrow K^+$$

such that

(i) $Sh(x, y) : J \rightarrow K^+$, for fixed $x \in J$, is a chain of connection

(ii) $Sh(0, y) = c_1$, $Sh(m, y) = c_2$ for some finite m

and (iii) $Sh(x, y)$ and $Sh(x + 1, y)$ are q-adjacent for $0 \leq x \leq (m-1)$.

The relation between chains of connection of "being q-shomotopic" is clearly reflexive, symmetric and transitive. Thus the finite set $\{c_i\}$ of all chains of connection in K fall into disjoint equivalence classes - which we naturally call the q-shomotopy classes.

A3.2 Face-saving maps

Given two complexes K, K' and an injective map

$$\beta : K \rightarrow K'$$

then it is of significance to know how β affects the q-nearness property. If $\sigma_r, \sigma_s \in K$ and if

(i) σ_r is q-near σ_s in K

implies (ii) $\beta(\sigma_r)$ is q-near $\beta(\sigma_s)$ in K' we shall say that β is a face-saving map (or an f-s map). We notice however that it does not follow that if σ_r is not q-near σ_s in K the $\beta(\sigma_r)$ is not q-near $\beta(\sigma_s)$ in K'.

Since a chain of q-connection $[\sigma_r, \sigma_s]_q$ is defined in terms of q-nearness we can see at once that $\beta[\sigma_r, \sigma_s]$ is also a q-connection in K'. It follows that

A face-saving map preserves q-connectivities

but A face-saving map need not preserve chain length

By the same considerations, referring to the definition of q-adjacent in section 3.0, it follows that

A face-saving map preserves the property of being q-adjacent

This means that, given a q-shomotopy Sh between the chains c_1 and c_2, there is a natural shomotopy induced between their images $\beta(c_1)$ and $\beta(c_2)$ by any face-saving map β. This induced shomotopy will be

$$Sh' : J \times J \to K'^{+}$$

and such that (i) $Sh'(x, y) = \beta \circ Sh(x, y)$ for fixed $x \in J$

(ii) $Sh'(0, y) = \beta(c_1)$; $Sh'(m', y) = \beta(c_2)$ for a

unique m' where the m' is defined by β, c_1, c_2 and Sh. It follows that

A face-saving map preserves q-shomotopy

A3.3 Loops in a complex

A chain of q-connection, $[\sigma_r, \sigma_r]$, which connects σ_r to itself and in which the only simplex which appears more than once is this σ_r, wil be called a q-loop, and denoted by

$$\ell = [\sigma_r, \sigma_r]_q \quad \text{or} \quad [\sigma_r]_q$$

If the length of the loop is (n - 1) we can regard ℓ as a mapping

$$\ell\{\sigma_r, \sigma_r, n\} = \ell\{\sigma_r, n\} = \ell : J \to K^+$$

Such a q-loop will be said to be based on σ_r, the base simplex. The particular q-loop of zero length is the simplex σ_r itself,

$$\ell_o = \ell\{\sigma_r, 1\} = \sigma_r$$

We also notice that the mapping ℓ induces an associated mapping ℓ_n, which also defines the loop, viz.,

$$\ell_n : J/_{nJ} \to K \quad K^+$$

Since $J/_{nJ} \cong J_n$, the additive group of integers (mod n), this map ℓ_n is such that, for example,

$$\ell_n(m\, n) = \sigma_r, \text{ for all } m \in J,$$

Thus ℓ_n "wraps" J around the loop an infinite number of times. If $\pi : J \to J_n$ is the canonical epimorphism, we can see that the following diagram is commutative,

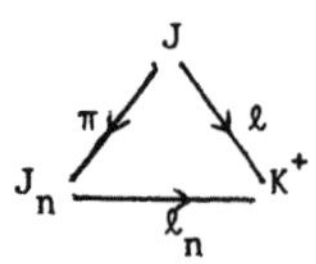

so that

$$\ell_n \circ \pi = \ell$$

Furthermore we can see that, for a fixed base simplex σ_r, the q-loops L_q can be given a <u>group structure</u> $\{L_q, *\}$. This is achieved by the following definitions.

<u>$\{L_q, *\}$</u>

(i) $\ell_o = \ell\{\sigma_r, 1\}$ is to be the group identity

(ii) the loop inverse to $\ell\{\sigma_r, n\}$ is to be the reverse loop $\bar{\ell}\{\sigma_r, n\}$

(iii) given $\ell_1\{\sigma_r, n_1\}$ and $\ell_2\{\sigma_r, n_2\}$ we define $\ell_1 * \ell_2$ as the loop

$\ell_3\{\sigma_r, n_1 + n_2\}$ with $\ell_3(i) = \ell_2(i)$ for

$0 \leq i \leq (n_2 - 1) = \ell_1(i - n_2)$ for $n_2 \leq i \leq (n_1 + n_2 - 1)$.

The meaning of (iii) is that the composite loop $\ell_2 = \ell_1 * \ell_2$ is the result of first mapping ℓ_2 and then ℓ_1, the continuations being performed through a q-face of σ_r.

<u>A3. 4 Coloops</u>

We define a q-coloop as an <u>homomorphism</u> from L_q into a coefficient group $\{G, \otimes\}$,

$$\lambda : \{L_q, *\} \to \{G, \otimes\}$$

This means that λ sends the base simplex σ_r to zero,

$$\lambda(\sigma_r) = \lambda(\ell_o) = 0$$

The set of all q-coloops associated with a particular base simplex σ_r forms a group under the usual composition,

$$(\lambda_1 * \lambda_2)(\ell) = \lambda_1(\ell) \otimes \lambda_2(\ell)$$

the identity map is λ_o and such that

$$\lambda_o(\ell) = 0_G \text{ for all } \ell \in L_q$$

and the inverse map λ^{-1} is such that

$$\lambda^{-1} * \lambda = \lambda * \lambda^{-1} = \lambda_o$$

We can denote this group of q-coloops by $\{L^q, *\}$. The generators of L^q will be a finite set of coloops $\{\mu_i\}$ such that

$$\mu_i(\hat{\ell}_j) = \delta_{ij} \quad \text{(Kronecker's delta)}$$

where the $\hat{\ell}_j$ are the generators of the group L_q. A typical generator $\hat{\ell}_j$ will be a q-loop $[\sigma_r, \sigma_r]_q$ in which σ_r appears initially and finally, but not elsewhere in the chain of connection. Thus $\hat{\ell}_j$ does not cross its own path. Any two such q-loops which are not identical must be distinct generators of L_q.

A3.5 The Shomotopy group Structure

We use the notion of shomotopy, introduced in section 3.1, to define an equivalence relation on L_q. This gives us a new quotient set S_q which possesses a natural group structure of its own. The idea behind this is to discard loops $\ell \in L_q$ which are sufficiently well q-connected with each other. This is likely to be significant for locally defined patterns - which will then presumably find local obstructions (only) between the elements of S_q. This is taken up in section A3.6 and in Section B.

An element $s \in S_q$ is to be the q-shomotopy class of q-loops in L_q. If ℓ is a representative of such a class then we write $s = [\ell]$. We obtain a group structure on S_q by defining a binary relation * in the following way.

(i) the identity element s_o is taken to be $[\ell_o]$

(ii) if $s_1 = [\ell_1]$, $s_2 = [\ell_2]$ S_q then * is defined by
$s_1 * s_2 = [\ell_1 * \ell_2]$

(iii) the inverse s^{-1} to s is then $s^{-1} = [\bar{\ell}]$ when $s = [\ell]$.

This gives us what will be called the qth shomotopy group

$$\{S_q, *\}$$

based on a s cific simplex σ_r in K.

But if we change the base simplex σ_r to σ_s, where σ_r and σ_s are in the same q-component, we can easily see that S_q is invariant up to isomorphism. In fact σ_r and σ_s must be connected by a chain, say $c\{\sigma_r, \sigma_s, m\}$, and if $\ell \in L_q$ based at σ_s, whilst $\ell' \in L_q$ based at σ_r, we have a correspondence by setting

$$\ell' = \bar{c} \circ \ell \circ c$$

But then ℓ' is q-adjacent to the unique ℓ'' where

$$\ell'' = \bar{c} \circ c \circ \ell$$

and so $[\ell'] = [\ell''] = [\bar{c} \circ c] * [\ell] = [\ell_0] * [\ell] = [\ell]$

giving the required isomorphism.

We thus see that the groups $\{S_q, *\}$ are characteristic of the q-components of K and give us insight into the local structure of each separate component.

Given a face-saving map $\beta : K \to K'$ then the qth shomotopy groups S_q will become the group S'_q. If $s = [\ell] \in S_q$ is mapped to $s' = \beta(s) = [\ell'] = [\beta\ell]$ it follows that if $[\ell_1] = [\ell_2]$ then $[\beta\ell_1] = [\beta^{-1}\ell_2']$. If this is the case, that is to say if

(i) $\beta : K \to K'$ is bijective

and (ii) β, β^{-1} are face-saving maps

then there is a (1-1) correspondence between S_q and S_q'. Since face-saving maps preserve the property of being q-shomotopic it follows that S_q and S_q' are isomorphic.

Under these circumstances, $S_q \cong S_q'$, we shall say that β is a <u>q-shomotopy equivalence</u>.

A3.6 Holes, objects and co-objects

If S_q possesses non-trivial elements (generators) each such $s = [\ell]$ will correspond to a weak kind of hole in the structure. This hole expresses the fact that $s \neq s_o$, that ℓ cannot be transformed to ℓ_o by q-adjacent loops.

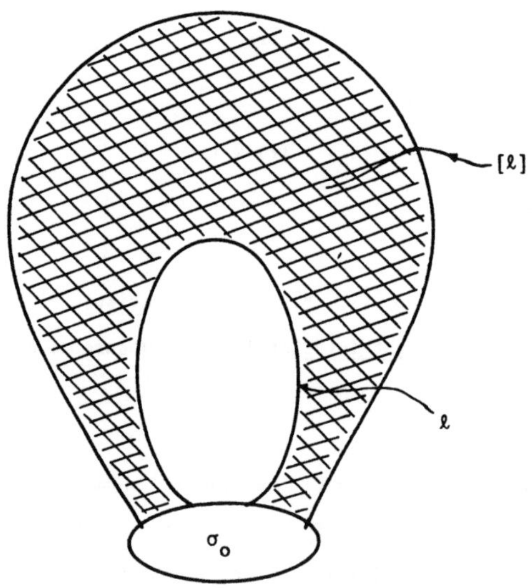

Figure 6. A local q-hole/[ℓ]

This means that a view of any one q-component of K which sees the local hole structure is naturally supplied by the group $\{S_q, *\}$. But the word "hole" is not quite appropriate in this context since, at $q = 3$, it is not necessarily an intuitive enclosure. In this respect we see that our q-shomotopy is not identical with the topologists homotopy.

The significance of the genuine "hole" in a structure such as physics has been discussed elsewhere[1] and it has been argued that such holes are indistinguishable from (solid) objects, observationally. But now that we have moved into a realm of Social Science and an abstract multi-dimensional space we must expect our "objects" to be somewhat less substantial - without necessarily losing their reality.

Accordingly we shall associate with each generator of the group S_q an entity which we shall call a local q-object. The nature of this entity, or object, will clearly depend on the specific complex K under consideration - but also must it depend on the possible coloop which expresses the local dynamics. As a "hole" the q-object is "not all there" (necessarily) but even so it possesses a reality of local structural significance.

It seems natural to associate with the q-objects mappings which shall characterise the dynamics "around" them, and to play a dual role - rather after the fashion in which coloops are dual to loops. We therefore introduce the notion of a co-object (or co-ob for short) as an homomorphism,

$$\zeta : \{S_q, *\} \to \{G, \otimes\}$$

Such a co-object must be such that it takes the value zero on any q-loop which is q-shomotopic to a (single) simplex.

Now it may well be that, having associated specific real "things" with the generators of S_q in some particular application, we can use the idea of "near-neighbour" to obtain a further relation on these things. By near-neighbour we could mean, for example, that the same q-adjacent loops which are used to defined (to "separate") the objects can be regarded as forming (in this new relation) suitable bridges between the objects.

Finally, we notice that two of our social q-objects which are distinct, need to be distinguishable at (q-1).

<u>A-References</u>

1.	Atkin, R.H.	Mathematical Structure in Human Affairs, Heinemann, London (1974)
2.	"	Research Report I (1972) - Urban Structure Research Project, University of Essex
3.	"	Research Report III (1973) - Urban Structure Research Project, University of Essex
4.	Dowker, C.H.	Homology Groups of Relations (1952) - Ann. Math. <u>56</u>, 84
5.	Hu, S-T.	Elements of General Topology, Holden-Day (1964)

B0.0 Introduction

This Section of the book constitutes an application of the structural language which has been developed and whose mathematical formulation (so far) is to be found in Section A. This particular application is to the University of Essex (Colchester, England), and is an attempt to show that the complex interrelationships, across the whole field of academic, social, political life of that community can be given a rational framework for its setting. But this framework is in the last analysis a mathematical one, showing in its development that the ambiguities of an idiomatic language, such as English, correspond to a well-defined structural property which is to be found in that mathematical analysis. The work was undertaken during 1973-4 as part of a Research Project financed by the Social Science Research Council, and the results have been published (1974) as a Research Report [v. B-reference (1)].

The method of attack has been to set up sets of data which are required for the analysis and to select those sets from various aspects of University life. The sets therefore refer not only to different categories, such as academic or social, but also inevitably refer to sets of people, such as students or staff etc.. These data are then available for analysis, usually by computer programs, and the results need then to be interpreted so as to emphasise the relation between the structural language and the intuitive experience of members of the community. The main aim of the research, and of the Report, is therefore the scholarly one of attempting to establish the relevance of the language as a vehicle suitable for a rational analysis of the community. It is not primarily concerned with criticising that community or in making specific recommendations about the way it should be organised or the way in which its members should behave.

A fundamental aspect of this research, and one which has been illustrated in previous Research Reports [v. A-reference 1,2,3] is the use of the idea of hierarchical levels N, (N+1), etc.. The word hierarchy, in this context, has a specific meaning which is well-defined in mathematical terms. It must not be confused with any other use of the word which might be associated with "class" or "caste". Nor is it of much significance as to which level is referred to as N-level - but the relation between successive levels is the important property. We can illustrate the idea in the following simple way by looking at sets of words describing (for example) some aspects of community life.

(N+1)-level	{Social Amenity}
N-level	{Catering, Sport}
(N-1)-level	{Food, Drink, Groceries, ..., Athletics, Ball Games, ...}
(N-2)-level	{Pies, Chips, ..., Beer, Wine, ..., Carrots, Milk, Meat, ..., Football, Tennis, High-Jump, Billiars, ...}

At each level we have a mathematical set which acts as a cover of the set below it. In general this cover is not a partition - in which the elements are disjoint. For example, at (N-1), "Drink" and "Groceries" probably share "Soft Drinks" at (N-2). It is because of this possibility of overlap that the use of the English words appears to be ambiguous. But in the mathematical sense this overlap can be given a precise expression via the ideas of q-connectivity and in that sense the "ambiguity" is removed - indeed it becomes an essential part of the overall structure of the language. In common speech it is clear that we do not keep the hierarchical levels distinct in any one sentence or statement. Presumably this gives us greater freedom of debate and possible misunderstanding.

If we let X be the set of things at (N-1) and Y be the set of things at N-level, the statement that Y acts as a cover of X merely means that the elements of Y are names for subsets of the set X and that no element of the set

X "escapes" coming under at least one element of the set Y. If in particular the cover is a partition then no element of the set X can be found in more than one element of the set Y, so the elements of Y are disjoint.

This idea of hierarchical levels is applied to all aspects of the study so that it refers to University Committees as well as to items of business which are discussed therein, and to individual people and sets of people who go to make up various groups etc.. It also refers to the physical buildings and their partitioning into sub-units, rooms, etc.. We find in fact that the hierarchical levels which are needed to discuss the whole structure are the following

(N+3) (N+2) (N+1) N (N-1) (N-2)

In these levels (N+3) refers to affairs which are outside the University community, but related to it whilst the remaining five levels are restricted to the internal structure. The overall picture of how these levels are related and typically constituted is shown in Figure 7.

If we reflect on the significance of the hierarchical levels for the individual member of the community we must notice that each person is presumably free to experience all of the levels. The level (N-2) is commonly referred to in the Report as the level of the individual. But we mean by this only the level at which the individual ceases to be part of a cover for any lower levels. His concerns at that level will therefore be personal concerns and these can only be introduced into the analysis by introducing lower levels (N-3) etc.. On the other hand when (e.g.) the Chairman of a Department attends a meeting and wears his "departmental hat" then he is operating in the structure as an N-level entity. When the Senate meets and deliberates it is fulfilling a role which is primarily at the (N+2)-level.- Senate being part of the cover which is defined on the (N+1)-level sets of committees.

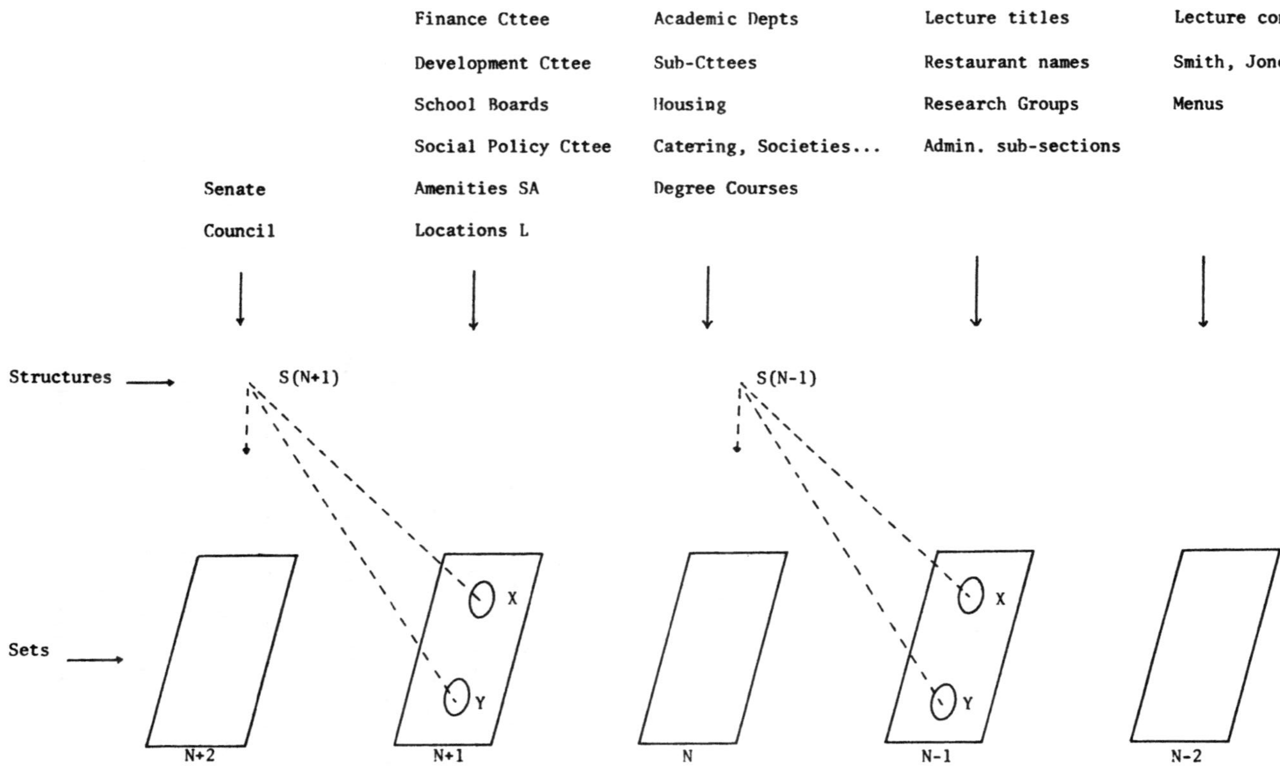

Figure 7. Hierarchical Structure

In a similar way the items of business which circulate around the University and through its committees and bars will be expressible in this hierarchical scheme - the details of these properties being pursued in later sections.

But at any one hierarchical level (e.g.) N, there exist mathematical relations between distinct sets at that level. These mathematical relations then become the subject of Q-analysis after being expressed in the form of a data file suitable for analysis by the computer programs. This Q-analysis gives two views of the activities which exist between the elements of two particular sets. It has been extensively illustrated in earlier Research Reports and in this Report the precise computer outputs of most of the data files are listed in the Appendix. The two views which the analysis gives are best described as global and local. In the global view we obtain a multi-dimensional structure (which is discussed in terms of generalised geometrical concepts) of the whole of the relation which is being considered. In broad terms it means that an abstract geometrical structure can be identified, and which we describe as the N-level backcloth S(N), as the actual "space" which must support all relevant action (such as spending money, altering priorities, making decisions, transmitting information, etc.). This global geometry is by no means uniformally the same in all directions and so it is essential to have a means of characterising it in terms of its peculiar connectivities. This is done largely by the consideration of what is called a structure vector $\underline{Q}$ and its associated obstruction vector $\hat{\underline{Q}}$.

The second view which the analysis gives us of S(N) concentrates on peculiarities of the local activities - as for example we need to notice that there are local hills and mountains near part of the earth's surface. The chief feature which the analysis throws up is that of q-holes - which turn out to be of local significance in any relation which is included in S(N). This strange kind of hole has been given the mathematical name of shomotopy [v. Section A3.5], and plays a role (not yet fully understood) in a mathematical group structure. In this study a large number of these local q-holes are discovered and the important thing for the language is to be able to see the significance of such local irregularities for various aspects of the community. Generally speaking it seems to be confirmed that action (of whatever kind) in the community can be seen as traffic in the abstract geometry and that this traffic must naturally avoid the holes (because it is impossible for any such action to exist in a hole). The holes therefore appear strangely as objects in the structure, as far as the traffic is concerned. Thus the words q-hole and q-object are used interchangeably throughout the discussion. The difference is a logical one in that the word "q-hole" describes a static feature of the geometry S(N) whilst the word "q-object" describes the experience of that hole by traffic which moves in S(N). Thus we can have a picture of this multi-dimensional activity in which the interplay of traffic and object can be illustrated, as in Figure 8.

The analysis seems to suggest that the real experience of the community by individual persons corresponds very closely with the global and local structure features to be found there. If there is any change in these features then the individuals experience them as strong pressures or forces which they think of as essentially social in their context. These structurally induced forces in the multi-dimensional space S(N) seem to transcend anything which the individual seems able to initiate in isolation in the community. Tampering with the structure, such as rearranging the committees or their membership or the division of responsibilities, has consequences which become far reaching throughout the structure. Perhaps we can suggest that most of these consequences might be predicted by monitoring the structure via the kind of analysis which is here outlined.

It is not surprising that one of the most interesting features of the

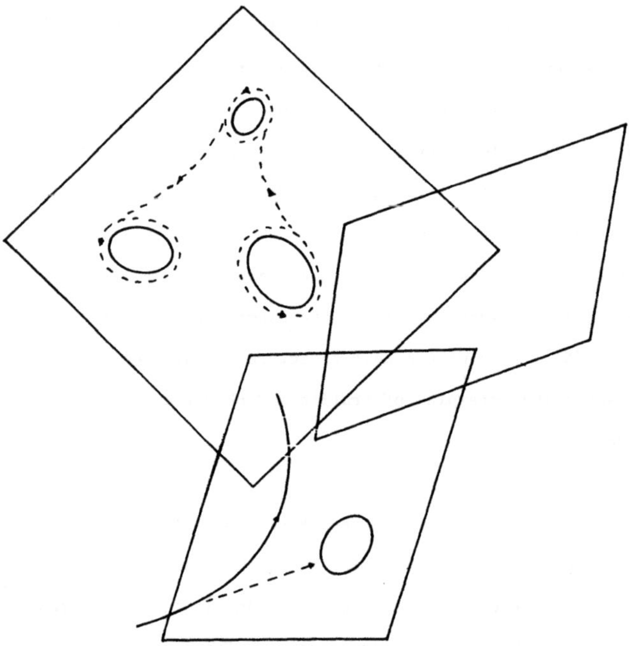

Figure 8. q-Objects/Holes in K(X)

University is its committee structure and the way this is involved in arriving at decisions on policies of all kinds. The analysis of this particular University from this point of view will be found in B2.2 to B2.4. The chief characteristic lies in a peculiar near-complete disconnection between the structures S(N+1) and S(N+2), on the one hand, and S(N+3), S(N), S(N-1), S(N-2) on the other. This can be represented schematically as follows

S(N+3)

S(N+2) S(N+1)

S(N)

S(N-1) S(N-2)

When this disconnection is playing a significant role in the action which is represented by relevant kinds of traffic in the structure there is a strong analogy with the idea of a medieval citadel. This relatively loose way of describing the structure is commonly used in subsequent sections. Although this citadel is only defined in the multi-dimensional space it is nevertheless a very real entity.

This idea of "what is real" and "what is abstract" needs to be reflected upon at some length. For it is particularly easy to imagine that when we refer to a multi-dimensional space we are referring to something which in some sense is not "real". This is no doubt due to a long history of education about what "dimensions" really represent. We are severely conditioned to think that the 3-dimensional space of the engineer has a reality which surpasses all other realities, and are inclined to ask the question "How can I imagine an n-dimensional space?" The answer is simply that it is unnecessary to imagine any such thing. These higher dimensional spaces are the spaces in which we already experience the complexity of relationships between ourselves and other objects, people, ideas, etc.. In other words we are already living in a multi-dimensional space - the 3-dimensions of the scientist are necessary but not sufficient for our intuitive experiences. The mathematician helps to make the idea of dimension precise in algebraic terms, and this study is an attempt to demonstrate that these algebraic "abstract" terms are merely another way of describing what we have always known intuitively.

B1.0 The Visual Backcloth

We begin by asking the question "What does the University fabric look like?", and we attempt to answer this by setting up a suitable hierarchical collection of visual sets. We refer to these sets in the following way:

Visual Features	Set	Hierarchical Level
Wide Views	W	N + 1
Views	V	N
Facade strips	F	N - 1
Details	D	N - 2

There are four elements in the set W and these can be described as views which are seen as one approaches the University from its four sides. We describe them as follows:

W1 - View from the Wivenhoe House approach route above the Visitors' carpark.

W2 - View from the South Side of the Boundary road in front of the two residential towers.

W3 - View from the river side of the Boundary road.

W4 - View from the north towers carpark.

In a similar way there are 20 elements in the set V, 36 elements in the set F and 84 elements in the set D. Descriptions of these vertices are illustrated in the tables on pages 203/7. A selection of photographs which illustrate some of the vertices are shown in this section.

By using this approach we are able to move through a hierarchy of visual features in which, for example, the vertices of V at the N-level act as a mathematical <u>cover</u> of the vertices of F at the (N-1)-level. The connectivity structure which the elements of a particular set enjoy is the structure which results from the details of this cover. We therefore have the following mathematical relations which completely describe the visual backcloth which exists in this hierarchy.

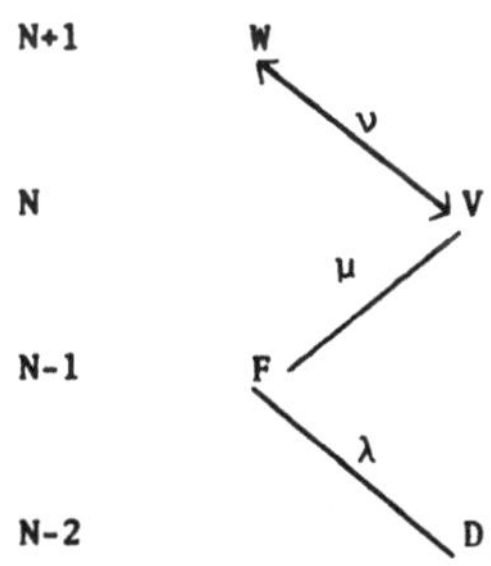

$$\nu \subset W \times V$$

$$\mu \subset V \times F$$

$$\lambda \subset F \times D$$

[Note: We do not discuss what would be a natural (N+2)-level - which we would expect to have one vertex only and that consisting of an aerial view of the whole campus].

A Q-analysis of these relations gives us the following structure vectors $\underline{Q}$.

Complex	Structure Vector $\underline{Q}$
$K_W(V)$	$\{\overset{5}{2}\ \ 2\ \ \overset{3}{1}\ \ 1\ \ 1\ \ \overset{0}{1}\}$
$K_V(W)$	$\{\overset{3}{1}\ \ 1\ \ 1\ \ \overset{0}{1}\}$
$K_V(F)$	$\{\overset{9}{1}\ \ 2\ \ 5\ \ 7\ \ \overset{5}{10}\ \ 8\ \ 5\ \ 3\ \ 5\ \ \overset{0}{1}\}$
$K_F(V)$	$\{\overset{8}{1}\ \ 3\ \ 3\ \ 6\ \ 5\ \ 3\ \ 5\ \ 5\ \ \overset{0}{1}\}$
$K_F(D)$	$\{\overset{11}{1}\ \ 1\ \ 6\ \ 8\ \ 9\ \ 12\ \ \overset{5}{12}\ \ 11\ \ 9\ \ 13\ \ 5\ \ \overset{0}{1}\}$
$K_D(F)$	$\{\overset{20}{1}\ldots 1\ \ 1\ \ 1\ \ \overset{8}{2}\ \ 2\ \ 2\ \ 4\ \ 5\ \ 4\ \ 7\ \ 6\ \ \overset{0}{1}\}$

These structure vectors give us a global view of the multi-dimensional geometry inherent in this visual backcloth and which is defined by the relations λ, μ, ν. The relation between W and V shows that the obstruction vectors $\hat{\underline{Q}}$ are very nearly zero, being respectively

$$\hat{\underline{Q}} = \{\overset{4}{1}\ \ 1\ \ 0\ \ 0\ \ 0\ \ 0\}$$

for $K_W(V)$, and

$$\hat{\underline{Q}} = \{0\ \ 0\ \ 0\ \ 0\}$$

for $K_V(W)$. This means that there is effectively zero obstruction to any change in visual emphasis as we move over the 20 vertices of V - seeing as we do their manifestation in the elements of the set W. A more coloquial way of putting this is to say that at the (N+1)-level the differences in the features are smoothed out in a visual sense. It is therefore at this level that we have a sense of the University as a visual unity.

But as we descend to the next hierarchical levels and consider the relation μ between V and F the striking change in the structure vectors show that obstruction to the shift of visual attention is far from zero. At the level of $q = 0$ there is indeed only one component. At this hierarchical level we can therefore say that we see the fabric as a unity only if we can be sensitive visually to one view (or one facade strip) at a time, If we are more visually sensitive to the elements in F or in V then there is a considerable obstruction as we try to shift our visual attention from one set member to another. For example, in $K_V(F)$ there are 10 components at q = 5. This means that there are 10 elements of the set V which do not share 6 elements of the set F. They therefore fall into distinct components and appear as distinct visual features to anyone who is sensitive at this level of q. Precisely, these 10 components consist of the following elements of V [c.f. page 203]:

[V1], [2], [V3], [V4], [V6, V10], [V7], [V9], [V15], [V16], [V20].

It will be seen that the views which describe the Squares come out as separate components. At this hierarchical level they therefore have individual and distinct visual contributions to make to the backcloth.

Referring to the conjugate complex $K_F(V)$ we notice that the maximum obstruction occurs once more at q = 5 when the number of components is 6. These are in the following:

[F1], [F2], [F3], [F6], [F12], [F14].

These can be identified by referring to the set elements on page 204. They appear, via this analysis, as the outstanding disconnected facade features which are related to the N-level views in F.

At the bottom of the hierarchical structure the relation λ between F and D gives us structure vectors in a much higher dimensional space. The obstruction to a change in visual emphasis is at its highest at q = 5 and q = 6 in $K_F(D)$, whilst this occurs at q = 2 in $K_D(F)$. In the latter case there are 7 components in $K_D(F)$ and these 7 components contain 2 which have more than 2 numbers in them. These 2 non-single components are as follows:

[F1, F2, F8, F10, F11, F12, F13, F15, F22, F23, F36, F4, F21, F24, F25], [F17, F18, F19, F20].

The second of these 2 components contains 4 identical simplices which are D20, D23, D67 and illustrate the visual distinctness of the Towers. The other component serves as a link among the Details of the valley buildings and can be identified by referring to page 204.

There is one component in $K_D(F)$, $q = 2$, which dominates all the others. This component contains the following details - vertices of the set D [D3, D4, D5, D10, D11, D21, D25, D65, D66, D8, D7, D36, D13, D14, D22, D24, D63, D81, D69]. This set can be identified from the list on page 205.

In $K_F(D)$ those Facade Strips which stand out among their neighbours are the ones with the greatest eccentricities. We select 6 of these to illustrate.

$\underline{K_F(D)}$

Ecc (F23) = 1	Ecc (F28) = 1
Ecc (F24) = 3	Ecc (F29) = 2
Ecc (F25) = 2	Ecc (F35) = 11

The outstanding facade with eccentricity of 11 is F35 which is Wivenhoe House. The next highest eccentricity occurs with F24 which is the Hexagon facade; F25 is the facade of the Library and this occurs with eccentricity = 2. In order to appreciate what a high eccentricity means we need to compare such a case with an example of the lowest eccentricity, viz zero. The commonest visual occurrence of zero eccentricity occurs with F5, F6, F7 - which we have designated as elements of the set F which referred to types of concrete wall.

If we use the eccentricity as a measure of what is visually noticeable in a local sense then the values which we have listed illustrate the intuitive awareness of Wivenhoe House as an outstanding feature on the campus and, to a lesser extent, the facade of the Library and the Hexagon restaurant.
In comparison with this the features which have zero eccentricity are those associated with the obvious finish of the concrete walls and structures in the valley. These eccentricities refer of course to the hierarchical level (N-1) and if we refer to the complex $K_D(F)$ we obtain a corresponding picture at the level of (N-2). The interesting eccentricities are the following:

$\underline{K_D(F)}$

Ecc (D25) = $^{10}/_{11}$ concrete strip

Ecc (D3) = 0 beam

Ecc (D21) = $^{1}/_{8}$ pillar

Ecc (D11) = $^{2}/_{5}$ medium panel

There can be no doubt that the lack of colour in the Details and the Facades make it more difficult to be visually sensitive to the high dimension of some of the elements. An increase in the use of colour, particularly for elements of the set D, would emphasise the dimension of these simplices in the structures $K_F(D)$ and $K_D(F)$. This would give a sense of unity and connectedness to the valley buildings. At the moment this sense of connectedness requires the sensitivity of an architect to be fully appreciated. It illustrates the rather obvious fact that the visual fabric of the University means vastly different things to different viewers. Since it is most obvious that most viewers will agree on low connectivity values, such as $q = 0, 1$, it is not surprising that the "sameness" of the visual features throughout the University becomes the commonest experience among those who work there. The only way in which the visual richness of the fabric can be appreciated is to make the dimensions of the simplices more visually obvious. The use of colour seems the obvious solution to this problem.

It is possible that those visually sensitive people who object to the use of colour do so because it seems to be making too obvious what is discernable already and in this sense can be criticised in their eyes as making the fabric visually vulgar. But for those people whose visual sensitivity to the simplex dimensions is that much lower - and this must surely include the majority of us - this criticism is not really understood, because it does not correspond to the visual experience.

To the unsophisticated viewer of the visual backcloth, whose sensitivity is stretched to the limit at $q = 2$ or 3, the University buildings are all connected at all hierarchical levels - and this means that the whole structure has a monotony which is unavoidable. The analysis shows that this is not the case at higher q-levels. The exception for the unsophisticated viewer is Wivenhoe House with its extraordinary eccentricity value of 11. But to appreciate this as a visual entity distinct from the structures in the valley requires only the crudest of sensitivities. If we cannot appreciate that Wivenhoe House is visually distinct from the valley buildings then we must be almost totally blind.

B1.1 Shomotopy Objects in the Visual Structure

The concept of the shomotopy groups $\{S_q, *\}$ has been introduced and discussed in Section A. Such a group S_q, for a particular q-value, gives us means of studying the local connectivity of a particular component in the backcloth. Whereas the structure vector $\underline{Q}$ helps us to appreciate the global connectivities so the generators of the groups S_q provide us with objects of attention to be found "inside" a particular q-component of S(N). This is because a generator of S_q defines a kind of hole in the abstract geometry and this hole must be impervious to any "traffic" which exists in a geometry. Thus the hole manifests itself to the traffic as if it were an object. Speaking loosely, we would say that the object offers a hard surface to the traffic through which it cannot penetrate; it must therefore bounce off in order to remain in the geometry which surrounds the hole.

This study of the University is particularly concerned with the occurrence and significance of generators (q-holes or q-objects) for the shomotopy groups $\{S_q, *\}$ associated with the backcloth structure. We begin to see the occurrence of these generators even in the visual backcloth discussed in B1.0. They are found by the use of a computer program called SHOM, written in Basys and run on the University's PDP 10.

In the complex $K_V(F)$ there are two q-objects at the level of $q = 2$. The first one is defined by the two-chain of connectivity

V1 V2 V4 V3 V1

whilst the second 2-object is given by the chain

V1 V2 V7 V6 V9 V1

If we consider the first generator and refer to the list of vertices in the set V (page 203) we see that it is made up of those views which are called Square 4, Square 3, Square 2 and Green court 1 - the latter referring to the enclosed area containing lawn and trees behind the Physics building. These physical components form themselves into a <u>loop</u> which is the <u>2-object</u>. They link themselves in pairs as shown in Figure 9 by sharing elements of the set F. They must be said to describe a <u>visual object</u> in this backcloth and in order to appreciate it a viewer needs to see each in turn on an imaginary journey through the University. Thus this visual object is an <u>entity</u> associated with the physical arrangements of the structure; it exists as a visual object which requires the exercise of a visual memory on such a journey. If no such object existed in the structure then there would be no identifiable entity in the

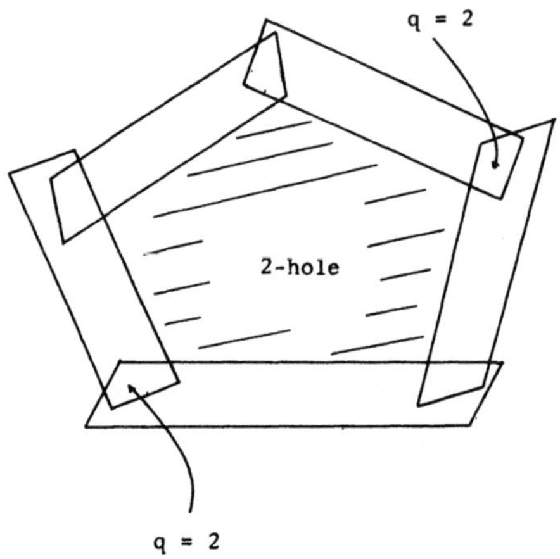

Figure 9. A 2-hole in the visual structure

memory. The presence of this visual object will induce "traffic" through the University's physical buildings in order to grasp the visual identity of the University. This identity, which remains in the visual memory, will presumably be a collection of visual objects such as are here defined.

The second 2-object is made up of the views of Square 4, Square 3, Green Court 2, Valley entrance and Square 5. The role which the Squares play in both of these visual objects would seem to be a justification for the original architectural plan and to indicate that they have been so designed as to produce visual objects by these appropriate combinations, at the level of q = 2. It is of interest to speculate as to the impressionable experience of a visual object at a higher q-level, but there are none in the complexes $K_V(F)$ or $K_F(V)$.

It is tempting to propose that the existence of a q-object in any structure must in itself give rise to traffic in that structure. In the case of a visual object that traffic can be simply the traffic of viewers. This would also imply that in some way not yet fully understood the <u>q-objects act as a source of energy for the possible traffic</u>. This theme will be explored throughout this study as we consider the existence of roles of the shomotopy generators occurring across the whole spectrum of the University backcloth.

B1.2 Social Amenity Backcloth S(N)

In order to study the social amenities of the University it was necessary to produce a set L of locations which adequately covered the physical buildings of the University. We regard this set L as an N-level set containing 30 elements. These 30 elements L1, ..., L30 can be identified by reference to the plates reproduced between pages 82 & 83. Together with this set L we take a set SA which describes at the N-level what might be called the social amenities offered by the University. There are 14 elements in SA as follows:

1	Retail	8	Hobby Clubs
2	Landscape	9	Indoor Games
3	Housing	10	Outdoor Games
4	Catering	11	Political Clubs
5	Services	12	Social Clubs
6	Academic Societies	13	Voluntary Service
7	Entertainment	14	Student Agencies

The Q-analyses for the relation between SA and L are given on pages 208, 209.

The structure vectors are the following:

<u>KY(X)</u>

$$\underset{\sim}{Q} = \{\overset{22}{1} \ldots \overset{14}{1} \; 2 \ldots \overset{10}{2} \; 3 \; 2 \; 2 \; 2 \; 1 \; 1 \; 1 \; 2 \; 1 \; \overset{0}{1}\}$$

<u>KX(Y)</u>

$$\underset{\sim}{Q} = \{\overset{9}{1} \; 1 \; 1 \; 2 \; 3 \; 5 \; 3 \; 1 \; 1 \; \overset{0}{1}\}$$

The simplex of maximum dimension in KY(X) is the one with the name Services. The structure vector for KY(X) can therefore be checked against the Q-analysis on page 208 and shows us how the various Social Amenities form themselves into connected components in the structure which represents this mathematical relation.

The values of Q_q for KY(X) are small enough to demonstrate that the obstruction vector $\hat{\underset{\sim}{Q}}$ is practically zero. This means that, provided we take the whole of the campus as equally available at any given moment, there is almost 0-obstruction to a change of pattern across the Social Amenities. This would seem to illustrate the commonly held belief that the physical connections in the University buildings make freely available all parts of the structure to the people who live and work there. But the weakness of this point of view is the weakness which derives from regarding the extensive University buildings as essentially compact at a point. In practice the different locations in the set L are not equally available or accessible to any member of the community but can only be enjoyed by extensive perambulations among the corridors and up and down the many staircases which connect the structure. However, with that proviso, each structure vector shows that in this global sense there is practically zero obstruction to the enjoyment of social amenities which are scattered throughout the buildings.

Details in set D / Wivenhoe House, (N-2)-level

View W3 in set W, (N+1)-level

View V2 in set V, N-level

plus vertices in set F, (N-1)-level

View V7 in set V, N-level

plus vertices in set F, (N-1)-level

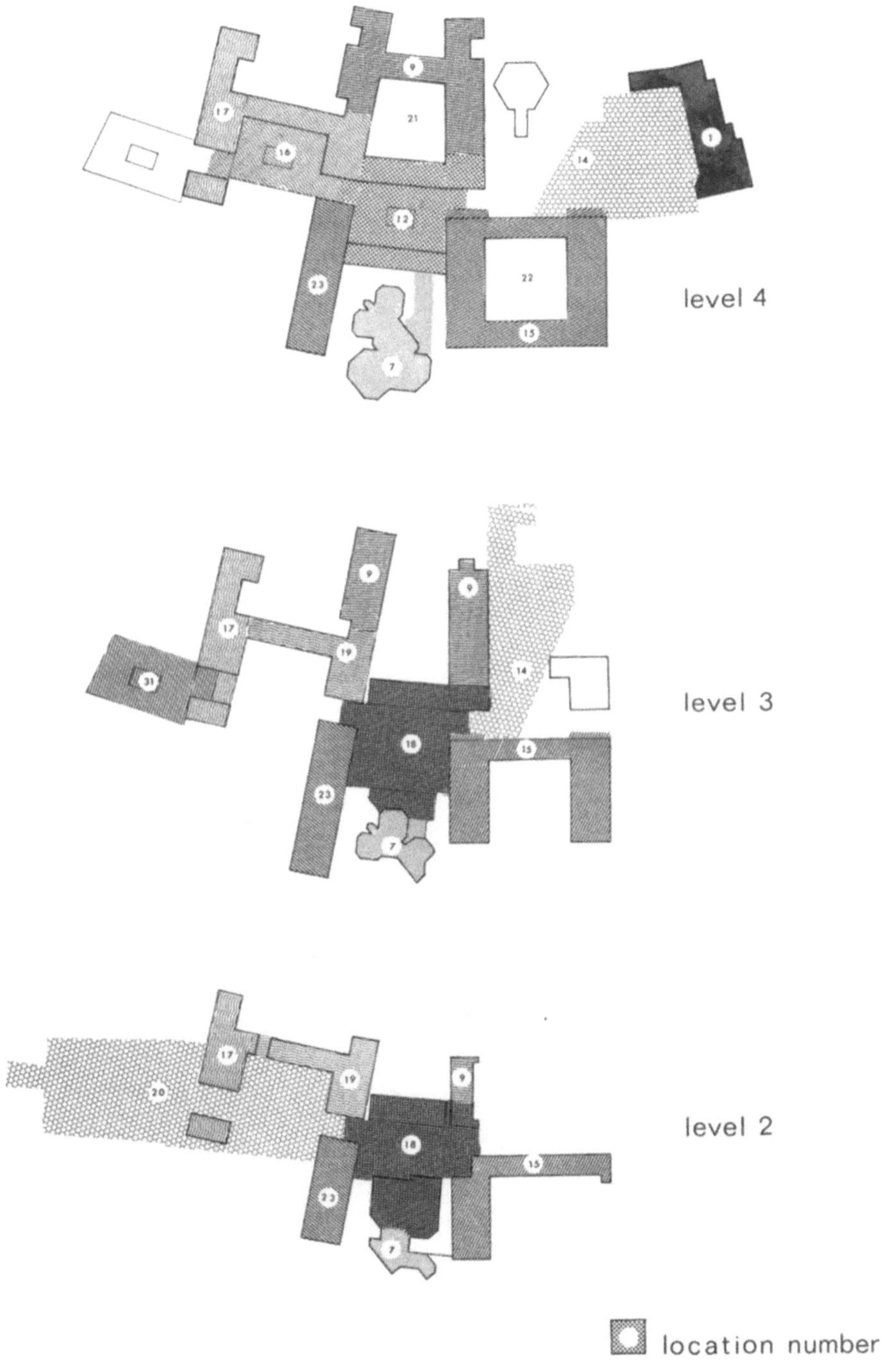
level 4
level 3
level 2
location number

If we turn to the conjugate complex KX(Y) which relates to the locations in L to the amenities in SA, we see that the structure vector in a similar way gives rise to a near zero value. This means that if we take, for example, a pattern π which gives us the distribution of people on any one occasion enjoying the various amenities in the various locations, then $\Delta\pi$ refers to a possible redistribution across the University buildings. The near zero obstruction vector in this complex indicates that such a change of pattern is not seriously obstructed. The reason for this is the same as that expressed in the previous paragraph and depends upon the same constraints in which an individual must disregard the physical problem of moving from one part of the buildings to another.

This is by no means a trivial problem, particularly for undergraduates who must spend most of their time on the campus searching for their own entertainment and leisure pursuits. The only geometry in which they can move to experience these amenities is that provided by this complex and the problem of a change in pattern for the student body as a whole is a problem which depends upon the connectivities of this complex.

We notice, for example, that at $q = 0$ in KX(Y) there are 28 locations in a single component. This means that there is a chain of locations which includes all 28 members and which is such that each neighbouring pair share one Social Amenity. An individual wishing to understand what Social Amenities are available at this N-level must be willing to traverse the whole University (excluding 2 areas of parkland) in order to do so. If we move to the level of $q = 4$ we see that there are five components in KX(Y) (see page 209) four of these are locations 1, 2, 11, 15 and the only component with more than one location in it is that consisting of the following:

(4 18 7 9 16 24 30)

This may be identified by referring to the plates between pages 82 & 83 .

B1.3 Shomotopy Objects in the Social Amenity Structure

In the complex $K_L(SA)$ there is one generator for the shomotopy group $\{S_3, *\}$ and this persists into the shomotopy group $\{S_2, *\}$. This 2-object, which is also a 3-object, is defined by a set of locations in the cover set L. Precisely it is given by the generator

$$\{L1 \quad L7 \quad L4 \quad L9 \quad L6 \quad L18 \quad L1\}$$

It will be seen from the plate diagrams that L1 is the area of Square 5 and the Library, L7 is the Lecture Theatre Block, L4 is the building at level 6 adjacent to the Lecture Theatre Block, L9 is the building which rises from level 2 to level 5 and commonly referred to as the Mathematics Building, L6 is what is commonly referred to as the Chemistry Building,level 5 and above, and L18 consists of the buildings beneath Square 3.

This generator defines a Social Amenity object at the levels of $q = 3$ and $q = 2$. It therefore constitutes a hole in the complex around which the amenity-searching traffic must flow. Because it is a 2-object it is not possible for any individual, in forming part of the traffic around it, to enjoy more than three amenities in any one location. Its existence corresponds to the intuitive feeling that the social centre of the University, in so far as it exists, is in reality a "run-around" the physical buildings in the valley. As we saw in a previous section, such an object generates the traffic and in this instance it is reasonable to see that the traffic is likely to consist of frustrated students trying to come to grips with the heart of the social life of the University.

Of course, it may be fairly objected that the complexes which derive from SA v L do not completely exhaust the social life of the University. But in order to meet this objection we can consider an extension of it so as to produce a backcloth S(N) which is more comprehensive. This is done by moving down the hierarchy from N-level to (N-1)-level and by considering the complexes at this lower level - which must naturally extend the N-level words in SA to suitable (N-1)-level words in other sets. In order to do this we consider in the next sections the structures which are to be found under the set names of Catering, Societies, Housing and Services.

The relations we shall consider will be those which are defined between the elements of these sets at (N-1)-level and the elements of the set L at N-level.

B14. Stress in the Catering Structure

This is the study of a relation at the (N-1)-level between the set of Catering establishments:

SQ3 = Square Three Restaurant
CB = Coffee Bar
HEX = Hexagon restaurant
MCR = Maths Common Room
WH = Wivenhoe House
VB = Valley Bar

and the set of Services

C/T = Coffee or Tea
SN = Snack
M = Meal
D = Alcoholic Drink

Snack S(N) is understood to mean the possibility of a substantial snack such as a selection of sandwiches and rolls, but not merely the availability of biscuits and sweets. Meal (M) is understood to mean a hot meal, commonly meat and two vegetables. Drink means the existence of a fully stocked licensed bar. This set of four kinds of catering services could be extended down to the (N-2)-level to include a detailed list of all possible purchases.

We studied the structure induced by the relation which gives the availability of different meals etc., in the various catering establishments and the geometry which resulted was itself a function of the price scales in the various restaurants. The data was collected during a week at mid-term in November 1973.

We have the following cost matrices, where the entries are values in pence.

Average cost

	CB	HEX	SQ3	WH	VB	MCR	
C/B	3	4	4	4		4	Coffee
SN	10	10	15	15	-	-	Bread & Cheese
M	-	31	36	42	-	-	Meat, 2 veg, sweet
D	-	-	14	14	13	-	Pint of beer

In order to appreciate what the structure means to a customer (whom we shall refer to as a student) we need to postulate the existence of a slicing vector which each student must use as his upper limit of expenditure on the four items C/T, SN, M, D. For example, the personal slicing vector

$$\{3 \quad 12 \quad 12 \quad 0\}$$

means that the student regards the upper limit of what he is willing to pay for each of the four items as 3p, 12p, 12p, 0p respectively. The 0p means that he does not take alcoholic drink. If we take this particular vector as a set of slicing parameters on the cost matrix we obtain the structure KY(X): CB = a 1-simplex $\langle$C/T, SN$\rangle$. This means that he must use the Coffee Bar and either takes coffee or tea or a snack. The structure of the catering amenity for this student is therefore reduced to a single 1-simplex to be found in a single establishment. On the other hand a student with a more generous slicing vector

{10 30 50 20}

finds that the catering structure consists of the following

<u>KY(X)</u>

WH, SQ3	= 3-simplex		$\langle$C/T, SN, M, D$\rangle$
HEX	= 1	"	$\langle$C/T, M$\rangle$
CB	= 1	"	$\langle$C/T, SN$\rangle$
MCR	= 0	"	$\langle$C/T$\rangle$
VB	= 0	"	$\langle$D$\rangle$

<u>KX(Y)</u>

C/T	= 4-simplex		$\langle$MCR, CB, SQ3, HEX, WH$\rangle$
SN	= 2	"	$\langle$CB, SQ3, WH$\rangle$
M	= 2	"	$\langle$HEX, SQ3, WH$\rangle$
D	= 2	"	$\langle$SQ3, VB, WH$\rangle$

The structure KX(Y) tells us the geometry in which the separate kinds of food and drink exist as polyhedra on the campus. For this affluent student the geometry is rich and the choice is varied. He may freely move about the campus retaining a reasonably wide choice for his selected meal as he does so. In comparison the poor student moves in a pitifully simple geometry which centres on the Coffee Bar and C/T, SN. In between these two cases we can expect to find the vast majority of students with their separate geometrical structures.

As we move from one such structure to another, consequent upon changing the slicing vector, we are in effect enforcing the change in the geometric backcloth which must carry the catering provision. This means that, as the geometry is so warped, the students experience t-forces which express any such change. These forces are the direct expression of changes in costs and/or changes in what any individual catering establishment provides. The personal experience of these forces may be imagined by contemplating that one individual - such as the reader - finds that his circumstances require him to change his slicing vector from the affluent one above to the poor one. The sudden shrinking of the abstract geometry directly corresponds with his sensation of adjusting his behaviour as a potential consumer in the catering establishments. The same effect will be produced quite naturally by a change in prices throughout the University assuming that the slicing vectors of the students remain constant. In modern conditions of inflation, it is highly likely that the slicing vectors will be steadily decreased by the individuals so that the overall effect of any increase in prices will be that much greater.

Another aspect of catering as an amenity, is the problem of being served at the time that is most convenient and the fact that a long queue will effectively distort the geometry for the customer. Thus we can study <u>queueing-time</u> and <u>queue-length</u> as patterns on the structure which act as stress induced by geometrical distortion. If the stress is too great in one part of the structure it can mean that a simplex must be removed by the customer. This can mean that he must experience the social force which drives him to move from one restaurant or bar to another in search of his meal. This social force, induced by a distortion of the geometry, is a painful experience; its effect is of the same nature as that produced by rising prices or falling slicing vectors.

The underlying geometry defined by the relation between the sets {SQ3, CB, VB, HEX, MCR, WH} and {C/T, SN, M, D} is, at various times of the day, as follows:

Time	KY(X)	$\mathcal{Q}$	KX(Y)	$\mathcal{Q}$
0945-1000	CB = $\langle$C/T, SN$\rangle$	$\{1, \overset{0}{1}\}$	SN, C/T = $\langle$CB$\rangle$	$\{\overset{0}{1}\}$
1000-1115	CB = $\langle$C/T, SN$\rangle$ MCR = $\langle$C/T$\rangle$	$\{1, \overset{0}{1}\}$	C/T = $\langle$CB,HEX,MCR$\rangle$ SN = $\langle$CB, HEX$\rangle$	$\{1\ 1\ \overset{0}{1}\}$
1115-1200	CB = $\langle$C/T, SN$\rangle$ VB = $\langle$D$\rangle$	$\{1, \overset{0}{2}\}$	C/T, SN = $\langle$CB$\rangle$ D = $\langle$VB$\rangle$	$\{\overset{0}{2}\}$
1200-1400	CB = $\langle$C/T, SN$\rangle$ HEX = $\langle$C/T, M$\rangle$ SQ3, WH = $\langle$C/T,SN,M,D$\rangle$ VB = $\langle$D$\rangle$	$\{1,1,1,\overset{0}{1}\}$	C/T = $\langle$CB,HEX,SQ3,WH$\rangle$ SN = $\langle$CB,SQ3,WH$\rangle$ M = $\langle$HEX, SQ3, WH$\rangle$ D = $\langle$VB, SQ3, WH$\rangle$	$\{1\ 1\ 1\ \overset{0}{1}\}$
1400-1430	CB = $\langle$C/T, SN$\rangle$ VB = $\langle$D$\rangle$	$\{1, \overset{0}{2}\}$	C/T, SN = $\langle$CB$\rangle$ D = $\langle$VB$\rangle$	$\{\overset{0}{2}\}$
1430-1500	CB = $\langle$C/T, SN$\rangle$	$\{1, \overset{0}{1}\}$	C/T, SN = $\langle$CB$\rangle$	$\{\overset{0}{1}\}$
1500-1600	CB, HEX = $\langle$C/T, SN$\rangle$ MCR = $\langle$C/T$\rangle$	$\{1, \overset{0}{1}\}$	C/T = $\langle$CB, HEX, MCR$\rangle$ SN = $\langle$CB, HEX$\rangle$	$\{1\ 1\ \overset{0}{1}\}$
1600-1700	CB = $\langle$C/T, SN$\rangle$	$\{1, \overset{0}{1}\}$	C/T, SN = $\langle$CB$\rangle$	$\{\overset{0}{1}\}$
1700-1800	CB = $\langle$C/T, SN$\rangle$ SQ3 = $\langle$C/T, M$\rangle$	$\{2, \overset{0}{1}\}$	C/T = $\langle$CB, SQ3$\rangle$ SN = $\langle$CB$\rangle$	$\{1\ \overset{0}{1}\}$
1800-1900	CB = $\langle$C/T, SN$\rangle$ VB = $\langle$D$\rangle$ WH = $\langle$C/T, M, D$\rangle$	$\{1, 2, \overset{0}{1}\}$	C/T = $\langle$CB, WH, SQ3$\rangle$ M = $\langle$WH, SQ3$\rangle$ D = $\langle$WH, VB$\rangle$ SN = $\langle$CB$\rangle$	$\{1\ 2\ \overset{0}{1}\}$
1900-2100	CB = $\langle$C/T, SN$\rangle$ VB = $\langle$D$\rangle$ WH = $\langle$C/T, M, D$\rangle$	$\{2, \overset{0}{1}\}$	C/T = $\langle$CB, WH$\rangle$ M = $\langle$WH$\rangle$ SN = $\langle$CB$\rangle$ D = $\langle$WH, VB$\rangle$	$\{2\ \overset{0}{1}\}$
2100-2230	CB = $\langle$C/T, SN$\rangle$ VB, WH = $\langle$D$\rangle$	$\{1, \overset{0}{2}\}$	C/T, SN = $\langle$CB$\rangle$ D = $\langle$VB, WH$\rangle$	$\{1\ \overset{0}{2}\}$

This underlying geometry overlooks distance, the over-riding consideration being that of the walk from the Valley establishments to Wivenhoe House (WH). For many people this would introduce a partition of the complexes KY(X) into two disjoint components, viz.,

{SQ3, MCR, HEX, CB, VB} and {WH}

The queueing-times and queue-lengths, for a typical day mid-week and mid-term (November 1973), are listed in the following table, at 15-minute intervals.

Time	CB		MCR		HEX		SQ3	
	Time	Length	Time	Length	Time	Length	Time	Length
0945	6	24						
1000	6	18			2	4		
1015	2	9	7	10	2	3		
1030	4	11	0	0	2	5		
1045	2	3	10	25	3	7		
1100	7	25	12	22	6	17		
1115	4	9			4	11		
1130	2	6						
1145	1	3						
1200	2	2			7	24	16	26
1215	4	9			3	7	8	12
1230	4	10			3	9	3	8
1245	5	16			3	8	2	8
1300	4	16			14	44	13	31
1315	7	23			8	24	10	26
1330	4	16			1	3	5	12
1345	3	12			0	0	1	2
1400	3	8			0	0	0	0
1500						9		
1515			2	4		9		
1530			1	4		15		
1545			2	4		2		
1600			4	14		0		
1700							11	30
1715							4	15
1730							2	6
1745							4	8
1800							6	12

Queue length is the number of people in the queue at the beginning of a 15-minute period; queue time with respect to time interval is the time it takes (in minutes) for the last person in the queue to finish being served.

Both the queue-time and queue-length constitute patterns on the structure KY(X) during any time interval of the day. These patterns can therefore induce stress in the same structures. By apportioning tolerance levels for these parameters we can study the distortion (or warping) of the geometry involved (in the Catering backcloth). Any such distortion is then a measure of the structural t-forces which exist throughout the day. These t-forces are those directly experienced by the people who are moving in the Catering backcloth (at the (N-1)-level).

If we set the queue-length tolerance parameter at the level of 20 (we join a queue if the number of people in it already is less than 20) we obtain a geometry for the period 1030-1115 as follows:

Time	KX(Y)	Q
1030	C/T, SN = $\langle CB, HEX \rangle$	$\{1 \quad \overset{0}{1}\}$
1045	C/T, SN = $\langle CB, HEX \rangle$	$\{1 \quad \overset{0}{1}\}$
1100	C/T, SN = $\langle HEX \rangle$	$\{\overset{0}{1}\}$
1115	C/T, SN = $\langle HEX, CB \rangle$	$\{1 \quad \overset{0}{1}\}$

This shows that, at various times, in this interval, there are the following t-forces acting in the structure:

A repulsive 2-force, evacuating the simplex $\langle MCR, CB, HEX \rangle$

A repulsive 1-force, evacuating the simplex $\langle MCR, CB \rangle$

A repulsive 1-force, evacuating the simplex $\langle MCR, HEX \rangle$

Looking for a cup of coffee during this time is likely therefore to subject one to any or all of these forces, due to the distortion introduced by the traffic which carries queue-length.

One result of this is an attempt by customers to by-pass the queue-length stress, by refusing to feel it. In the HEXagon restaurant this has resulted in an informal arrangement whereby customers ignore the SNacks queue and go straight to the cash desk with their cups of coffee. The result has been (February 1974) that the Catering Committee has requested that this arrangement be made formal by building a new C/T service bar in the HEX. They are therefore bending before the t-forces which the warping of the geometry has introduced into the structure.

If we slice the queue-time stress by a standard slicing vector

{10 10 10 10}

for the services {C/T, SN, M, D} - which means that the tolerance waiting time is 10 minutes for any service - then we obtain a typically distorted geometry like the following

Time	KX(Y)	Q	Underlying Q
1245-1315	SN = ⟨WH, SQ3⟩ C/T = ⟨HEX, SQ3⟩ M = ⟨WH⟩ D = ⟨VB, SQ3, WH⟩	$\{\overset{2}{1}\ \ 2\ \ \overset{0}{1}\}$	$\{\overset{3}{1}\ \ 1\ \ 1\ \ \overset{0}{1}\}$

This means the presence of

a 3-force of repulsion, evacuating ⟨CB, HEX, SQ3, WH⟩

a 2-force of repulsion, evacuating ⟨CB, HEX, WH⟩

a 2-force of repulsion, evacuating ⟨HEX, SQ3, WH⟩

a 2-force of repulsion, evacuating ⟨SQ3, WH, CB⟩

The only way of reducing these forces is to have a higher tolerance with respect to queue-time, for example, a slicing vector of

{20 20 20 20}

The existence of these t-forces is the cause of people constituting traffic around the Catering Establishment at various times of the day.

By monitoring these changes in the geometry, day by day and week by week, we could clearly identify the t-forces which are relevant to the search for refreshment on the campus. Making the problems well-defined in this mathematical sense is a necessary condition to their control.

B1.5 (N-1) v. (N-2) Housing Structure

We obtain more insight into the N-level amenity Housing by examining the relation between the (N-1) and (N-2)-level cover sets. These vertices are constructed from the published document, "University of Essex, Student Accommodation 1973-4" and from facts and interpretation arrived at in discussion with the University's Housing Officer.

Some comparison between the years 1972-3 and 1973-4 is given since new accommodation became available in October 1973, which produced non-trivial change in the structure. The vertices are listed with descriptions below:

(N-1)-vertices

Contract-1 Contract-2 Contract-3 Contract-4	Contract lodings. Students live with Landlady on Bed & Breakfast basis
House-Clacton-A House-Clacton-B House-Clacton-C House-Clacton-D	University owned property in Clacton divided into bedsitting rooms with shared facilities.
House-Colchester-A House-Colchester-B	University owned property in Colchester divided into bedsitting rooms with shared facilities.

*West-End-A *West-End-B *West-End-C *West-End-D *West-End-E	University owned property in Colchester divided into bedsitting rooms with shared facilities.
Tower Flat-A -	Flat on Campus in Rayleigh, Keynes, Tawney or William Morris Tower.
Tower Flat-B -	Flat on Campus in Eddington or Bretrand Russell Towers.
*Avon Way -	Flat in University Development on Greenstead Estate.
G/M1 G/M2 G/M3 G/M4 G/M5 G/M6	Flats for graduate and married students in one of the towers on Campus
*G/M7 -	As above, but flat in Avon Way development
G/M8 G/M9	University owned house divided into two flats for two student families.
Tower Flat - Staff-A Tower Flat - Staff B	Staff Flat in towers on Campus
Wivenhoe House-A Wivenhoe House-B	Two flatlets in Wivenhoe House for day by day bed & breakfast (short term).
S.H. - A S.H. - B S.H. - C S.H. - D	Stanstead Hall is on loan to the University and is divided into four flats for staff. Residents have the communal use of some of the large rooms such as the hall and dining room although their flats are self-contained.

* not available 1972-3.

<u>(N-2)-vertices</u>

1.	K-WORKING	KW	WORKING KITCHEN
2.	K-AREA	KA	KITCHEN AREA
3.	K/DINER	KD	KITCHEN DINER
4.	K/DINER(S)	KDX	KITCHEN DINER-SHARED
5.	K(S)	KX	KITCHEN SHARED
6.	LIVROOM(S)	LRX	LIVINGROOM-SHARED
7.	DINROOM	D	DININGROOM
8.	DINROOM(S)	DX	DININGROOM-SHARED
9.	LIVROOM/STUDY	LS	LIVINGROOM-STUDY
10.	STUDY	S	STUDY
11.	STUDY(S)	SX	STUDY-SHARED
12	HALL	H	ENTRANCE HALL
13.	BATHSHOWER	B	BATH OR SHOWER ALWAYS AVAILABLE
14.	BATHSHOWER(S)	BX	DITTO-SHARED
15.	W.C.	T	TOILET

16.	W.C.(S)	TX	TOILET-SHARED
17.	LIMITED BATHS	BL	BATHS LIMITED TO 2 PER WEEK
18.	DOM SERV	DS	DOMESTIC SERVICE
19.	1/2 DOM SERV	HDS	PUBLIC AREAS CLEANED
20.	BREAKFAST	BF	BREAKFAST PROVIDED
21.	EVENING DRINK	ED	HOT EVENING DRINK PROVIDED
22.	E/G METERED	PM	ELECTRICITY, GAS, LIGHTING, HEATING EXCLUDED
23.	E/G $^{1}/2$ METERED	PHM	SOME LIGHTING OR HEATING INCLUDED
24.	E/G INC	PI	ELECTRICITY, GAS, LIGHTING, HEATING INCLUDED
25.	LINEN	LI	LINEN PROVIDED AND LAUNDERED
26.	STUD/BED-1	SB1	STUDY BEDROOM GRADE 1
27.	STUD-BED-1(S)	SB1X	DITTO - SHARED
28.	STUD/BED-2	SB2	STUDY BEDROOM GRADE 2
29.	STUD/BED-2(S)	SB2X	STUDY BEDROOM GRADE 2-SHARED
30.	BEDSIT-1	BS1	BEDSITTER GRADE 1
31.	BEDSIT-1(S)	BS1X	BEDSITTER GRADE 1 - SHARED
32.	BEDSIT-2	BS2	BEDSITTER GRADE 2
33.	BEDSIT-3	BS3	BEDSITTER GRADE 3
34.	BEDSIT-4	BS4	BEDSITTER GRADE 4
35.	BEDROOM	BR	BEDROOM
36.	BEDROOM-D	BRD	DOUBLE BEDROOM
37.	UNFURNISHED	U	UNFURNISHED
38.	LOUNGE	LO	LOUNGE
39.	LIVROOM	LR	LIVINGROOM
40.	PLAYROOM	PL	PLAYROOM
41.	LAUNDRYROOM	LA	SHARED LAUNDRY ROOM
42.	BEDSIT-1-D	BS1D	BEDSITTER GRADE 1 DOUBLE
43.	BEDSIT-2-D	BS2D	BEDSITTER GRADE 2 DOUBLE
44.	BEDSIT-3-D	BS3D	BEDSITTER GRADE 3 DOUBLE
45.	BEDSIT-4-D	BS4D	BEDSITTER GRADE 4 DOUBLE

The first column is the set of computer output names.
The second column is the set of abbreviated vertex names, X denotes 'shared'.

If we use the notation

X = set of (N-2) vertices, Y = set of (N-1) vertices

we obtain the structures KY(X) and KX(Y) for the two years 1972-3 and 1973-4. The corresponding structure vectors are:

<u>1972-3</u>

KY(X) $\underline{Q} = \{\overset{14}{2}\ \ 3\ \ 1\ \ 2\ \ 1\ \ 1\ \ 7\ \ \overset{7}{10}\ \ 10\ \ 7\ \ 5\ \ 2\ \ 1\ \ 1\ \ \overset{0}{1}\}$

KX(Y) $\underline{Q} = \{\overset{16}{1}\ \ 2\ \ 3\ \ 2\ \ 3\ \ 3\ \ 2\ \ 2\ \ 3\ \ 3\ \ \overset{6}{2}\ \ 1\ \ 1\ \ 1\ \ 1\ \ 1\ \ \overset{0}{1}\}$

<u>1973-4</u>

KY(X) $\underline{Q} = \{\overset{14}{2}\ \ 3\ \ 1\ \ 2\ \ 1\ \ 1\ \ 7\ \ \overset{7}{15}\ \ 12\ \ 8\ \ 5\ \ 1\ \ 1\ \ 1\ \ \overset{0}{1}\}$

KX(Y) $\underline{Q} = \{\overset{20}{1}\ \ 2\ \ 3\ \ 3\ \ 3\ \ 3\ \ \overset{14}{3}\ \ 4\ \ 2\ \ 2\ \ 3\ \ 3\ \ 4\ \ 2\ \ 1\ .\ .\ .\ \overset{0}{1}\}$

These vectors reflect the changes introduced into the structure at the beginning of the academic year 1973-4 and marked in the list Y above.

The vector $\underline{Q}$ for KX(Y) shows a little change, as an obstruction, but the maximum dimension has risen from $q = 16$ to $q = 20$. This is a natural consequence of there being more houses/flats available over which the elements of X can be distributed.

The significance changes in the $\underline{Q}$ for KY(X) occur for the values of Q_q at $q = 5, 6, 7$. Thus Q_7 changes from 10 to 15, as the properties in West End (Colchester) enter the structure at $q = 7$. This means that, in this part of the Social Amenity backcloth there have been introduced

t-forces of attraction

at $t = 7, 6, 5$. Individuals express this t-force simply as an inclination to live in the properties concerned. But also, of course, the increases in Q_7, Q_6, Q_5 mean that the obstruction to changes in π^7, π^6, π^5 on this structure has also increased. Thus, people who live in those properties because of the q-values involved are less likely to want to move or to be able to move without some reappraisal of their ranking of these q-values.

....

If we look at the local structures KY(X), KX(Y) we find the following "q-objects" in the Housing context.

<u>1972-3</u>

KY(X) at $q = 4$:

{G/M2 TOWER FLAT-STAFF A G/M8 S.H.-D WIVENHOE H.-A S.H.-C G/M2}

<u>1973-4</u>

KX(Y) at $q = 5$:

{KW B DX H LS T KW}

and {KW T DX LRX PM H B KW}

The first is a 4-object (or "hole") which can be loosely described as encompassing properties which form "self-contained high quality accommodation". This is an identifiable object in KY(X); the traffic which it engenders, which circles round it, consists of all those tenants who are searching for that ideal. Those which occupy any property in the generator above are "on the edge" of that object. Those in property which lie further away (but shomotopically related to the generator) are that much more remote. The best that a would-be tenant can do is to move into the generator listed above. Of course there is another kind of traffic in the structure, and that consists of the owners (private and university) who administer the property and who build it, maintain it, and set the standards. This traffic has in effect created this 4-object.

The object "disappears" at $q = 3$; a tenant can therefore not see that the traffic is sustaining the object only if he "drops his sights" to $q = 3$ (at the most).

The 5-objects in the conjugate complex KX(Y) are generated by sets of (N-2)-vertices. Any two neighbours in the list share a 5-face whose 6 vertices are to be found in the (N-1)-set Y. Such an object is an expression of the fact that nowhere in the Structure can one find a single property which possesses all the vertices in the list. Searching for such a property makes one part of that traffic in KX(Y) which cycles round these 5-objects. The 5-object can only be removed (the 5-hole filled) by creating new properties of a much higher standard. This would introduce higher order t-forces into the structure, diverting the traffic to occupancy. The consequences for costing and financing would be functionally dependent on that set of circumstances. This might suggest that a case could be made out for an institution, such as the University, to ensure that at least one q-object should exist in its Housing Structure - either in KY(X) or KX(Y) or both. If, in contrast, it tried to eliminate any such objects (which would presumably reduce administrative traffic around them) then it would seem to be inviting difficult financial problems into the picture.

B1.6 A Rents Pattern Generator

When the simplicial structure for Housing is established, as in the previous section, it is possible to consider various patterns $\{\pi\}$ on it. These could be Demand (numbers on the waiting list for each type of accommodation), Maintenance Costs, Population in occupancy, or Rents to be charged. We shall illustrate the idea of such a pattern by considering the example of Rents.

The underlying geometrical structure is that of the previous section, viz., the relation between the (N-1)-level cover set and the (N-2)-level cover set. Since the matrix of data relating one level to the other is a weighted relation we have essentially 3 structures - obtained by slicing at SV = 1, 2, 3.

The exercise is an important illustration of the pattern generator (v. section A2.7).

$$e_o^{\Delta} \equiv 1 + \frac{\Delta}{1!} + \frac{\Delta^2}{2!} + \ldots + \frac{\Delta^r}{r!} + \ldots$$

operating on a zero-order pattern π^0. The use of this generator ensures that the zero-order pattern π^0 (giving values on the vertices at (N-2)-level) is equivalent without bias to higher graded patterns π^t throughout the whole structure. It therefore ensures "fairness" across the structure - in the sense that "you pay for what you get". The latter "things" constitute the (N-2)-vertices, and if the resulting rents are not acceptable to all the parties concerned it means only that they are in disagreement about this (N-2)-level set vertices. But this defines the area of discussion for them - which should not be expressed in N-level or even (N-1)-level terms.

The zero-order pattern used here is made up as follows:

$$\pi^0 = \pi^0(SV = 1) + \pi^0(SV = 2) + \pi^0(SV = 3)$$

We must also understand that if there are differences of opinion about the (N-2)-level vertices because, for example, a single bedroom in house A is "not of the same quality" as one in B, then this can only be settled by going to an (N-3)-level - in order to distinguish the two cases. There is no particular difficulty in then setting up a π^0 at (N-3) and using the generator e_o^{Δ} to produce rents at the (N-1)-level.

The following table of values of π^0 on the (N-2)-vertices is based on a norm which regards the student towers flats, the contract lodgings, the student Avon Way accommodation, and the University houses (Clacton and Colchester) - excluding the recent West End houses in Colchester - as an acceptable basis.

(N-2)-vertex	π° (pence/week)	(N-2)-vertex	π° (pence/week)
KW	60	PI	40
KA	30	LI	40
KD	50	SB1	240
KDX	25	SB1X	130
KX	30	SB2	218
LRX	20	SB2X	115
D	40	BL1	200
DX	20	BS1X	190
LS	40	BS2	180
S	40	BS3	156
SX	20	BS4	70
H	10	BR	250
B	40	BRD	396
BX	25	L1	-100 (minus)
T	20	LO	40
TX	5	LR	40
BL	13	PL	20
DS	200	LA	20
HDS	30	BS1D	450
BF	117	BS2D	416
ED	10	BS3D	364
PM	0	BS4D	270
PHM	20		

This π° gives 15 exact agreements with present rents (1973-4), 18 to within 5p, and 21 to within 25p of present rents. The complete set of values of rents are generated by e_0^Δ and identified by

$$\pi^p = e_0^\Delta (\pi^\circ) + \pi^*$$

where p is the dimension of the simplex under consideration and where the pattern π^* takes into account the extra simplex [(N-1)-level] of being self-contained. A simplex with the face

$$\langle \text{KW}, \ \text{B}, \ \text{T}, \ \text{BR}, \ \text{L/S} \rangle$$

is regarded as self-contained, whilst one with the face

$\langle$K/DX, BX, TX, BS1$\rangle$

is not. We have taken a constant value for π^* of 100p (a relative value which can be adjusted to taste). (See Table on next page).

We notice that the Wivenhoe House accommodation, which is not intended for occupation other than overnight stay, achieves th is by being heavily charged on a weekly basis.

But the other accommodation whose rents differ significantly from those generated by π^* are

Stanstead Hall A, C	(over)
Stanstead Hall B, D	(under)
GM 9	(over)
West End A, B, C, D, E	(over)
Towerflats (Staff) A, B	(over)

B1.7 Services and Societies v. Locations

In studying the way in which the various student societies use the physical Locations we set up a data file which relates the 73 societies with the 30 elements in L (this data refers to the academic year 1972/3). We use the word "society" to include any group of students which is officially recognised as such by the Student Union. It therefore covers various sporting groups, political organisations, and academic and specialist interests. The 73 vertices form a set at the (N-1)-level and if we use the notation

$$X = \text{Locations } L_i, \quad Y = \text{names of societies}$$

we obtain the two conjugate complexes KY(X) and KX(Y). The structure vectors

(N-1)-name as a p-simplex	p	Present weekly rent 1973-4	Generated Weekly rent	Difference
Contract - 1	8	527	527	0
Contract - 2	8	493	493	0
Contract - 3	8	417	417	0
Contract - 4	8	390	390	0
Clacton House A	7	370	370	0
Clacton House B	7	350	350	0
Clacton-House C	7	326	326	0
Clacton House D	7	240	240	0
Colchester Hse A	6	370	370	0
Colchester Hse B	6	320	320	0
Towerflat A	7	415	410	-5
Towerflat B	6	415	410	-5
GM 1	5	646	646	0
GM 2	5	676	676	0
GM 3	4	646	646	0
GM 4	4	504	510	+6
GM 5	4	480	480	0
GM 6	6	502	514	+12
GM 7	6	510	534	+14
GM 8	6	920	920	0
GM 9	6	720	666	-54
West End A	7	525	360	-165
West End B	7	475	338	-137
West End C	7	750	570	-180
West End D	7	700	536	-164
West End E	7	650	484	-166
Avon Way	6	415	410	-5
Wivenhoe House A	8	1960	871	-1089
Wivenhoe House B	8	1155	725	-430
Towerflat Staff A	7	900	726	-174
Towerflat Staff B	6	800	676	-124
Stanstead Hall A	14	1120	1250	+130
Stanstead Hall B	14	740	646	-94
Stanstead Hall C	13	800	946	-146
Stanstead Hall D	11	1600	1526	-74

which emerge from the Q-analysis are as follows

$$KY(X) : \underset{\sim}{Q} = \{\overset{3}{5} \;\; 8 \;\; 11 \;\; \overset{0}{3}\}$$

$$KX(Y) : \underset{\sim}{Q} = \{\overset{27}{1} \ldots \overset{11}{2} \;\; 2 \;\; 3 \;\; 4 \;\; 4 \;\; 4 \;\; 4 \;\; 6 \;\; 9 \;\; 9 \;\; 7 \;\; \overset{0}{3}\}$$

The first structure vector shows that the University is divided into three disjoint components at q = 0. One of these components contains 20 elements from the set L whilst the other two contain one each. This means that the vast majority of the societies meet in an area which covers practically the whole of the University being located in fact in those elements of L which are numbered.

1, 7, 11, 16, 18, 4, 6, 9, 24, 3, 5, 12, 14, 15, 17, 27, 28, 29, 2.

The whole of the analysis KX(Y) is displayed on page 218. From that analysis it will be seen that location L7, the Lecture Theatre Block, is the dominant centre for student society activities. After that the next most popular area of the University is L9, around Green Court I, as shown on the plate diagrams.

The dominant obstruction vector associated with KX(Y) is

$$\hat{\underset{\sim}{Q}} = \{\overset{11}{1} \;\; 1 \;\; 2 \;\; 3 \;\; 3 \;\; 3 \;\; 3 \;\; 5 \;\; 7 \;\; 7 \;\; 5 \;\; \overset{0}{0}\}$$

This has significantly high values at q = 1, 2, 3, 4 and this means that any student who wishes to explore the University in such a way as to retain the option of joining in at least 2 different societies in any one place finds that the University offers him a significant obstruction vector. The highest values occur at q = 2, 3 when the components of obstruction are each equal to 7. This means that the student must see the University buildings as constituting a large number of disconnected pieces in this backcloth of KX(Y). It is therefore quite impossible for him to identify society activity with one, or even two, pieces of the University. The obstruction inherent in $\hat{\underset{\sim}{Q}}$ is that which makes it impossible for him to change his mind about the use of societies as he moves round the University. As usual, it means that there is a relatively high degree of rigidity in the structure from this point of view.

This situation is reflected in the conjugate complex KY(X) where the effective obstruction vector becomes

$$\hat{\underline{Q}} = \{\overset{3}{4} \quad 7 \quad 10 \quad \overset{\bullet}{0}\}$$

This means thataat q = 1 the obstruction to changing one's mind from one society to another, in any one place in the University, has its maximum value of 10. It reflects the same situation as we have already noticed in KY(X). There is a high degree of rigidity in the structure which imposes the large obstruction to the change of any pattern π (such as attendance at the various society meetings) which might be defined on the structure.

......

The analysis of the shomotopy objects in the relation (see page 201) shows that there are two generators in KX(Y) at the level of q = 0. These are

G1 : {L1 L11 L3 L9 L7 L1}

G2 : {L4 L6 L9 L3 L7 L24 L15 L4}

These 0-objects cannot be undermined, in the sense that there is no value of q less than 0 which can provide us with simplices (locations in the University buildings) which can fill the holes. There are therefore two distinct holes which can be experienced by moving around the University locations in the loops denoted by G1 and G2. These holes, or objects, therefore generate the traffic of students moving around the University and looking for an option of taking part in the activities of a number of societies. The first generator describes pieces of the University which are either around Green Court I, at various levels, or in the Lecture Theatre Block. The components of G2 can be identified by referring to the plate diagrams.

The existence of these objects in the structure again illustrates

(a) the tendency for objects to initiate and generate traffic of all kinds in the structure.

(b) the intuitive experience of students who wish to use a piece of the backcloth in such a way as to allow them to be sensitive to a high q-value.

The fact that the objects are 0-objects means that this experience of not being able to find one's way to the centre of activity (where the "action" can be found) cannot be resolved, there being no lower q-values. It is not surprising that most students, particularly when they are new to the community, will describe this experience as one of frustration. In this sense the existence of the 0-objects might well be regarded as a bad thing.

....

The (N-1)-vertices which are in the set called Services are 21 in number and listed as follows:

LAUNDERETTE	TOWERS OFFICE
PARKING	HOUSING OFFICE
NURSERY	STUDENTS' COUNCIL
BANK	SECURITY
POST OFFICE	THEATRE
T.V. ROOM	CONCERT
GAMES ROOM	CAREERS
DANCE HALL	EXHIBITION
COMMON ROOM	NIGHTLINE
LIBRARY	DRI-PLA AREA
HEALTH SERVICE	

Denoting this set by Y and the set of locations by X we obtain the following structure vectors

$$KY(X) : \underline{Q} = \{\overset{12}{1} \ldots \overset{5}{2}\ \ 3\ \ 4\ \ 7\ \ 3\ \ \overset{0}{1}\}$$

$$KX(Y) : \underline{Q} = \{\overset{4}{3}\ \ 7\ \ 8\ \ 5\ \ \overset{0}{1}\}$$

The structure KX(Y) is very similar to the KY(X) of Locations v. Societies. The obstruction is high for $q \geq 1$ and the interpretation is on exactly similar lines. The University presents the feeling of being highly disconnected with respect to these particular services. It is intriguing that this reality is superficially counter-intuitive since the buildings are physically connected into one piece. But this illustrates forcibly the difference between that "space" which we really function in and that "space" which is 3-dimensional and is the one in which we erect our buildings.

It is not surprising that we find a 0-object under the shomotopy analysis. Its generator is the following

G1 : {L1 L7 L2 L15 L12 L3 L2 L4 L1}

Although these locations are distinct areas of the University, ranging from the Library to the Lecture Theatre Block, the analysis shows that they form a 0-object with all its attendant traffic-generation for members of the University.

B2.0 Forces experienced by Staff in a changing S(N)

We consider the distribution of academic and administrative staff throughout the University departments during the four academic years 1970/1 to 1973/4. This gives us a structure at the N-level and the changes in S(N) from this point of view gives us an expression of the personal experiences of Staff during that period. We study the changes in the structure in the following way.

Suppose we have two data files for successive years, denoting them by K_r and K_{r+1} and let $K_{r,\ r+1}$ be the structure which is obtained by subtracting the rth data file from the $r+1^{th}$ data file. For example, take $r = 0$ so that K_0

is the structure for 1970/1, K_1 is the structure for 1971/2, the K_{01} is the structure for the <u>increase</u> from the 1970 file to the 1971 file whilst K_{10} is the structure for the <u>decrease</u> from 1970/1. Reflection shows that the following relations then exist between these complexes

$$\underline{Q}(K_1) \leq \underline{Q}(K_0) + \underline{Q}(K_{01})$$

$$\text{Max } q \text{ in } K_1 \leq \text{Max } q \text{ in } K_0 + \text{Max } q.\text{in } K_{01}$$

During this four year period there has been a maximum of 32 departments in the University and these include both academic and non-academic. The list of departments is as follows:

ACADEMIC (ADMINISTRATION)
ART
CATERING
COMPUTING CENTRE
CHEMISTRY
COMPARATIVE STUDIESSSCHOOL
DATA BANK
DEAN OF STUDENTS OFFICE
ECONOMICS
ELECTRICAL ENGINEERING
ESTATES & PLANNING OFFICE
EDUCATIONAL TECHNOLOGY
FINANCE
GOVERNMENT
HOUSING
HISTORY
HEXAGON RESTAURANT
LANGUAGE CENTRE
LIBRARY
LITERATURE
MATHEMATICS
HEALTH CENTRE
MAINTENANCE
MATHEMATICAL STUDIES BOARD
PRINTING CENTRE
PHYSICS
PHYSICAL SCIENCE BOARD
SOCIOLOGY
SECURITY
SOCIAL STUDIES BOARD
STUDENTS COUNCIL
WIVENHOE HOUSE

There are 28 non-academic administrative grades (excluding that of Registrar) which have the following grade names:

MRGD	CLPR1	LJLA	CECO
MRGS	CLPS	TCHB	CJRO
MRGA	CLSS	TCHA	SNCW
MRAS	CLCS	TECO	SNMN
MRAJ	CLJS	TSNO	SNMD
MRAA	LSLA	CCHA	SNCH
CLPR2	LLAA	CSNO	SNSP

These grades fall into broad categories ranging from Senior Administrative (MR..) through Clerical (CL..), Technical and Specialist Administrative and Security personnel.

The distribution of these posts throughout the University was used to create data files of Grades v. Departments and the respective Q-analyses are listed on pages 219/23. The following structures were thereby obtained.

<u>1970/1</u> K_0

KY(X) $\underline{Q} = \{\overset{23}{1}\ 1\ 1\ 2\ 2\ 2\ \overset{17}{1}\ 2\ 2\ 1\ 1\ 1\ 1\ 1\ 2\ 1\ 2\ 2\ 3\ 1\ 1\ 1\ 3\ \overset{0}{2}\}$

KX(Y) $\underline{Q} = \{\overset{6}{6}\ 5\ 4\ 1\ 4\ 1\ \overset{0}{1}\}$

The first structure vector shows that for all practical purposes there was a 0-obstruction in KY(X) for that year. This 0-obstruction meant that as far as administrative jobs were concerned the whole of the University structure was equally accessible. It was possible to feel that the opportunities for being appointed in any particular administrative grade were roughly equal across the whole University. This was equivalent to zero-obstruction to the movement of potential applicants in the Administration. The highest q-value was 23 and this corresponded to Clerical Grade Secretary. The group of Clerical Senior Secretary followed a close second at q = 20 and in the Senior Administration the highest q-value occurred for Senior Administrative Assistant and Administrative Assistant, both at q = 7.

The second structure vector gives us a view from the departments towards the Administrative grades and this shows that the obstruction was certainly non-zero. The 6 components at $q = 6$ contained only 2 academic departments namely the Computing Centre and Electrical Engineering. The rest included Academic Administration, Finance and Maintenance. This meant that these were the departments where the most varied career prospects were offered to would-be employees. The spread of posts in the non-academic departments covered the top end of the administration whilst the spread which accounted for the Computing Centre and Electrical Engineering being in this position referred more to the variety of technical staff.

We now list the structure vectors which describe the changes in the complexes, using the notation $K_{r,\ r+1}$ as indicated. We notice that K_{01} indicates the increase from K_0 to K_1 whilst K_{10} denotes the decrease from 0 to 1.

<u>1970/1 - 1971/2</u>

$KY(X) : \underline{Q}\ (K_{01}) = \{\overset{11}{2} \ldots \overset{5}{3}\ 5\ 6\ 5\ 1\ \overset{0}{2}\}$

$KY(X) : \underline{Q}\ (K_{10}) = \{\overset{5}{1}\ 3\ 6\ 8\ 7\ \overset{0}{1}\}$

$KX(Y) : \underline{Q}\ (K_{01}) = \{\overset{5}{5}\ 3\ 8\ 9\ 5\ \overset{0}{1}\}$

$KX(Y) : \underline{Q}\ (K_{10}) = \{\overset{3}{2}\ 5\ 9\ \overset{0}{1}\}$

<u>1971/2 - 1972/3</u>

$KY(X) : \underline{Q}\ (K_{12}) = \{\overset{12}{1} \ldots \overset{7}{3}\ 3\ 3\ 4\ 5\ 5\ 3\ \overset{0}{2}\}$

$KY(X) : \underline{Q}\ (K_{21}) = \{\overset{8}{1}\ 1\ 1\ 2\ 3\ 6\ 6\ 2\ \overset{0}{2}\}$

$KX(Y) : \underline{Q}\ (K_{12}) = \{\overset{6}{1}\ 1\ 1\ 3\ 10\ 4\ \overset{0}{2}\}$

$KX(Y) : \underline{Q}\ (K_{21}) = \{\overset{5}{1}\ 1\ 1\ 4\ 7\ \overset{0}{2}\}$

<u>1972/3 - 1973/4</u>

$KY(X) : \underline{Q}\ (K_{23}) = \{\overset{16}{1} \ldots \overset{9}{2}\ 2\ 2\ 2\ 3\ 1\ 2\ 4\ 2\ \overset{0}{3}\}$

$KY(X) : \underline{Q}\ (K_{32}) = \{\overset{7}{1}\ 1\ 1\ 4\ 4\ 5\ 6\ \overset{0}{1}\}$

$KX(Y) : \underline{Q} = \{\overset{4}{3} \quad 6 \quad 9 \quad 6 \quad \overset{0}{3}\}$

$KX(Y) : \underline{Q} = \{\overset{2}{3} \quad 8 \quad \overset{0}{1}\}$

If we look at the complex KY(X) for K_{01}, K_{12}, K_{23} we obtain an interpretation of the forces experienced by staff who are contemplating a change of job. In each of the time intervals of one year the structure vector gives a multi-dimensional measure of the sense of job opportunity. Between 1970/1 and 1971/2 the greatest attractive force induced Clerical Grade Senior Secretary and Clerical Grade Secretary at $q = 11$ - that is to say throughout 12 departments. This persisted down to $q = 5$. The grade of Senior Administrative Assistant did not appear until $q = 3$, so opportunities came in only 4 departments. Parallel with these job opportunities we obtain, from K_{10}, some idea of the introduction of the lack of opportunities. The greatest lack of opportunity occurred in 6 departments for Clerical Junior Secretary and following this, in 5 departments, Administrative Assistant. As we move from K_{01} to K_{10} at $q = 2$ we see that the component changes from 5 to 8. This has a strange result for, at this q-level, lack-of-job-opportunity exceeded job-opportunity. The grades over which both of these forces operated were Clerical and Senior Administrative. This complicated shift in the potential career structure must be related to the changes in KX(Y), between K_{01} and K_{10}. The departmental expansions correspond to $\underline{Q}(K_{01})$ and at $q = 5$ the 3 components in the vector were in fact the Computing Centre, Finance and Maintenance. By comparison there occurred at $q = 3$ the contraction which affected the 2 components in $\underline{Q}(K_{10})$, namely, the Computing Centre and the Library. The departments which produced the expansion at $q = 4$ fell into 8 components and contained the 3 academic departments of Computing Centre, Language Centre, Data Bank - the other 5 being administrative. The contractions of the departments with respect to job grades, at the level of $q = 2$, gave 5 components, namely, Computing Centre, Chemistry, Finance, Library and Maintenance.

If the job opportunities which appear to an individual, or to a class of individuals, are to represent the possibilities of individual aspirations, then the changes in the structure (such as K_{01}) become in their turn an indication of the forces which act upon these aspirations. These are once more examined by what we have called t-forces and the changes which are manifest in the structure vectors (e.g. $\underline{Q}(K_{01})$) give us an indication of the values which t may take. These various t-forces form a complicated pattern which a class of job-seeking individuals may expect to experience. It seems plausible therefore to suppose that if we look at the vectors $\underline{Q}$ which describe the departments, for both K_{01} and K_{10}, then the associated obstruction vectors will indicate a measure of the rigidity which has been introduced into the system. This rigidity describes the attitudes of individuals for their aspirations towards obtaining a post in any one of the departments. If there is a high component in such an obstruction vector (for example, the number 8 at $q = 2$ in $\hat{Q}(K_{01})$) then <u>changes</u> in departmental oriented aspirations of job seekers are that much more inhibited. Thus if such high values keep recurring year after year there will develop an intuitive feeling among would-be Administrative staff that the University job opportunities have become increasingly channelled into disjoint pieces and that therefore an individual must increasingly feel obliged to work in one particular area rather than feel free to move about over the whole structure. This means, for example, that Administrative staff begin to identify themselves with narrow departmental interests; this is the meaning of the increasing rigidity described above.

This tendency is born out by the structure vectors $Q(K_{12})$, $Q(K_{23})$ where the values at $q = 2$ are 10 and 9 respectively. It is balanced to some extent, but not entirely, by a corresponding rigidity associated with departmental contractions with respect to grades. These are represented by the vectors Q for K_{10}, K_{21}, K_{32} - for KX(Y) in each case. Here the largest obstruction occurs at $q = 1$, with values 8, 6, 7.

The decreases in the structure correspond to Administrative grades being reduced in variety across the departments and, in the conjugate sense departments being reduced in variety with respect to Administrative grades.

In the year 1971/2 - 1972/3 the increase for the grades was most manifest at the level of Clerical Senior Secretary (compare the previous change). This change had a q-value of 12. The grades of Administrative Assistant and Senior Assistant Registrar were manifest at q = 2 in comparison. The decreases in this year occurred with Clerical Secretary at q = 8 and Administrative Assistant and Senior Administrative Assistant at q = 3. Moving to the conjugate complexes for this year we notice the increases occurring with q = 6 for the Computing Centre, with a corresponding increase in variety of administrative posts. Also, for example, at q = 2 there are 10 distinct components of which only the Computing Centre and the Library can be called academic. The decreases for the departments occur for Maintenance at q = 5, and for example, Catering, Finance, Language Centre, Maintenance at q = 2. The corresponding obstruction vectors show a reasonably high rigidity so that the departments concerned were showing a convergence to particular types of Administrative posts during this period. We notice, for example, that there was a tendency for the Senior Secretary establishment to be increased and for the "middle" Administrative grades to be decreased.

In the period 1972/3 - 1973/4 the leading increase was Clerical once more with Secretary occurring at q = 16 and Senior Secretary entering at q = 9. These grades dominated the whole University since other grades, for example, Administrative Assistant, did not enter the structure until q = 3. The decrease in variety occurred at a lower q-value of 7, being led by Clerical Senior Secretary and followed by Administrative Assistant at q = 4. The conjugate picture for this period showed an increase at q = 4 for the departments Library, Maintenance, Wivenhoe House whilst at q = 3 the departments concerned were Computing Centre, Finance, Government in addition. The decrease from the department point of view occurred at q = 2 for Academic Administrative, Computing Centre, Finance. An overall description of this analysis consists in saying that the expansion of the University at the N-level (expansion within the departments) has naturally resulted in the appearance of large t-forces (t taking values from 0-16) acting on classes of people who are seeking administrative careers. These t-forces are positive and attractive, helping to encourage aspirations in potential candidates. At the same time there has been a steady contraction in the structure and this contraction has resulted in another set of repulsive t-forces (t taking values from 0-8) which restricted variety of administrative job across the many departments. The individual seeking an administrative career in the University during the past 4 years will have had to respond to these t-forces. They will have influenced his choice of which jobs to apply for, what promotion he could reasonably aspire to, and which areas in the University he would feel able to work in. In the main there has been a repulsive force centering on the "middle" areas of Administrative Assistant so that it could well be that decisions taken in previous years by this grade will now be taken increasingly by the Clerical Principal Grade.

The final comment we need to make about the administrative grades is based on the shomotopy analysis of the data files. This has given us two 4-objects in the structure. The first occurs in KY(X) for the year 1972/3 and is an object generated by the following grades

MRGS CLPS TSNO TECO CLCS MRGS

This grouping of grades defines an administrative object whose elements are linked by sharing at least 5 departments. It indicates a traffic generator for administrative matters which need to flow through these departments. It is of interest that there is a significant gap between the Senior Assistant

Registrar grade and the other Clerical and Technical grades which define this object. This almost certainly means that these are the grades which generate most of the administrative traffic in the University.

The second generator is to be found in KX(Y) for 1973/4 at q = 4. It is a 4-object defined by

ACADEMIC ADMINISTRATIVE, FINANCE, MAINTENANCE, COMPUTING CENTRE, ACADEMIC ADMINISTRATIVE

This 4-object has elements which are linked by at least 5 administrative grades. It is therefore a traffic generator for would-be administrative personnel. To be more crude, it means that, taking into account the whole range of administrative posts from Technician to Senior Administrator, these 4-objects in the University form an object which offers the best chances of promotion for this class of person. Since the Computing Centre is present in this object largely because of the variety of Technician and Clerical Staff that it employs (and these occur at the "lower" end of the administrative scale) it follows that for the ambitious administrator in the University his main object of attention should be the 3 departments, Finance, Maintenance and Academic Administrative. In that most ambitious employees would sense this instinctively, it is hardly surprising; what is most important in this study is to bring out the point that such an intuitive appraisal of the situation corresponds exactly with the mathematical analysis inherent in the study of shomotopy objects.

....

We can examine the growth of the University during this same period from the point of view of the academic staff, using data files with the prefix AG v D. The Q-analyses for the 4 years are given on pages 224/5 and are based on a set of nine academic grades which are

PROFESSOR
READER
SENIOR LECTURER
LECTURER
5 GRADES OF RESEARCH STAFF

The following table describes the changes which correspond to those for the administrative staff

<u>1970/1</u>

$KY(X) : \underset{\sim}{Q}(K_0) = \{\overset{15}{1} \ldots \overset{10}{2} \; 1 \ldots \overset{4}{2} \; 2 \; 1 \; 1 \; \overset{0}{1}\}$

$KX(Y) : \underset{\sim}{Q}(K_0) = \{\overset{7}{1} \ldots \overset{0}{1}\}$

<u>1970/1 - 1971/2</u>

$KY(X) : \underset{\sim}{Q}(K_{01}) = \{\overset{7}{1} \; 1 \; 2 \; 5 \; 3 \; 1 \; 1 \; \overset{0}{1}\}$

$KY(X) : \underset{\sim}{Q}(K_{10}) = \{\overset{5}{1} \; 1 \; 2 \; 2 \; 5 \; \overset{0}{3}\}$

$KX(Y) : \underset{\sim}{Q}(K_{01}) = \{\overset{5}{2} \; 3 \; 3 \; 2 \; 1 \; \overset{0}{1}\}$

$KX(Y) : \underset{\sim}{Q}(K_{10}) = \{\overset{2}{2} \; 4 \; \overset{0}{3}\}$

<u>1971/2 - 1972/3</u>

$KY(X) : \underset{\sim}{Q}(K_{12}) = \{\overset{6}{1} \; 3 \; 4 \; 3 \; 3 \; 2 \; \overset{0}{1}\}$

$KY(X) : \underset{\sim}{Q}(K_{21}) = \{\overset{3}{4} \; 2 \; 4 \; \overset{0}{1}\}$

$KX(Y) : \underset{\sim}{Q}(K_{12}) = \{\overset{5}{1} \; 3 \; 2 \; 1 \; 3 \; \overset{0}{1}\}$

$KX(Y) : \underset{\sim}{Q}(K_{21}) = \{\overset{3}{2} \; 2 \; 4 \; \overset{0}{1}\}$

<u>1972/3 - 1973/4</u>

$KY(X) : \underset{\sim}{Q}(K_{23}) = \{\overset{5}{1} \; 3 \; 5 \; 6 \; 1 \; \overset{0}{1}\}$

$KY(X) : \underset{\sim}{Q}(K_{32}) = \{\overset{4}{1} \; 1 \; 4 \; 6 \; \overset{0}{1}\}$

$KX(Y) : \underset{\sim}{Q}(K_{23}) = \{\overset{3}{2} \; 5 \; 3 \; \overset{0}{1}\}$

$KX(Y) : \underset{\sim}{Q}(K_{32}) = \{\overset{2}{1} \; 6 \; \overset{0}{1}\}$

In 1970/1 the obstruction to a change of pattern on either the grades or the departments was effectively zero. There were Professors in 16 departments, Lecturers in 11, Senior Lecturers in 10 and Readers in 5. The departments with the largest variety of posts were the Computing Centre and Chemistry, at $q = 7$. This was no doubt due to the variety of Research staff as well as the "normal" academic staff. The forces which act upon the job seeking aspirations of would-be academic staff can be seen from the structure vectors for the changes in successive years. Thus the largest t-forces of attraction occurred at $t = 7, 6, 5$ for the successive years 1971/2, 1972/3, 1973/4. These corresponded to the grades of Lecturer, Reader, Lecturer. Thus in the year 1971/2 - 1972/3 there was an upsurge in the number of departments appointing Readers. The decreases corresponded to Research Assistant in 1971/2 - 1972/3 at $q = 3$, and Reader in 1972/3 - 1973/4 at $q = 4$. Thus whilst some departments were creating posts of Reader and/or Lecturer and/or Research Assistant others were reducing these numbers at some time during this period. The t-forces involved would therefore direct would-be applicants into different departments of the University in a natural way corresponding to this.

If we look at the conjugate complexes we obtain a view of the departments as related to the academic posts. In 1970/1 there is 0-obstruction to patterns across the departments. The subsequent changes are described by the structure vectors for KX(Y) in each successive year and for K_{01}, K_{10} etc.. The maximum t-forces corresponding to increases occurred at $t = 5, 5$ and 3 whilst the maximum t-forces corresponding to decreases occurred at $t = 2, 2, 2$. In 1970/1 - 1971/2 the 5-forces occurred in the Computing Centre and the Language Centre whilst in the next year the 5-forces occurred in the Mathematics Department (this corresponded to the setting up of a Fluid Mechanics Research Institute in that department, introducing a larger variety of staff), whilst in 1972/3 - 1973/4 the 3-forces occurred in the Computing Centre and Sociology. During the same years the t-forces of repulsion occurred in Chemistry, Mathematics, Language Centre, Literature and Physics.

Most of the changes for increase or decrease in this structure corresponded to the rise and fall in the provision for Research Officers of various kinds. But in departments which are either traditional-experimental or are needing to use Computer facilities extensively these changes are a reflection of a rise and fall of research work by the more permanent staff.

These changes, mirrored in the above table of structure vectors, therefore correspond to profound academic activity in the University. Since any proposed changes can be monitored by the Q-analysis right across the University it would seem desirable to discuss such proposals in this context and using this language.

There are no shomotopy objects in this part of the backcloth S(N). This means that there are no identifiable objects which generate traffic for job seekers, nor are there any objects which generate traffic around the departments (traffic in "business" which must be conducted between the grades).

B2.1 Structural significance of Degree Schemes

The degree schemes (providing the choice for undergraduate students) are the responsibility of Senate at the (N+2)-level, of the Schools at the (N+1)-level, and (informally) of the Departments at the N-level. The lecture courses and classes which constitute the degrees proper are to be found at the (N-1)-level, whilst the personal participation (of both students and staff) as well as the fine detail of specific lectures, etc. are found at the (N-2)-level. Thus a degree scheme (for a specific "first degree") is manifest at all the hierarchical levels within the university structure. It therefore plays a role in each of the backcloth structures S(N-2), ..., S(N+2). Furthermore, when the qualification, of having been awarded a degree, is taken "outside" into society at-large it is then manifest at the (N+3)-level.

We can study some of these effects by an analysis of two basic data files, viz., that of Schools v. Degree Schemes and of Departments v. Degree Schemes, and by following these through the 5 academic years from 1970/1 to 1974/5.

The following tables show the global structure vectors for each year for the data file "Schools v. Degree Schemes", together with information about the dimensions of the respective Schools-of-Study, etc..

<u>Data file</u>: Schools v. Degree Schemes (Y = Schools, X = Degree Schemes)

Year	$\underline{Q}$ for KY(X)	$\hat{q}$(Comp-St)	$\hat{q}$(Soc-St)	$\hat{q}$(Maths-St)	$\hat{q}$(Phys-Sc)
1970/1	$\{\overset{13}{1} \ldots \overset{7}{1}\ 2\ 1\ 2 \ldots \overset{0}{2}\}$	4	7	13	6
1971/2	$\{\overset{15}{1} \ldots \overset{8}{1}\ 2\ 1\ 2 \ldots \overset{0}{2}\}$	5	8	15	7
1972/3	$\{\overset{17}{1} \ldots \overset{10}{1}\ 2\ 2 \ldots \overset{0}{2}\}$	8	8	17	9
1973/4	$\{\overset{22}{1} \ldots \overset{11}{1}\ 2\ 1\ \overset{8}{2} \ldots \overset{0}{2}\}$	8	9	22	10
1974/5	$\{\overset{22}{1} \ldots \overset{14}{1}\ 3\ 3\ 3\ \overset{10}{4}\ 3\ 2 \ldots \overset{0}{2}\}$	10	10	22	13

Year	$\underline{Q}$ for KX(Y)	Total number of Degree Schemes
1970/1	$\{\overset{2}{1}\ 1\ \overset{0}{2}\}$	20
1971/2	$\{\overset{2}{1}\ 1\ \overset{0}{2}\}$	23
1972/3	$\{\overset{2}{1}\ 1\ \overset{0}{2}\}$	28
1973/4	$\{\overset{1}{1}\ 1\ \overset{0}{2}\}$	33
1974/5	$\{\overset{1}{2}\ \overset{0}{2}\}$	43

We immediately notice that $Q_0 = 2$ in each year, so that the University falls into <u>two disconnected academic components</u>. One of these is the School of Comparative Studieswwhilst the other consists of the remaining 3 schools, viz., Maths-Studies, Social-Studies, Physical-Sciences. This corresponds to a partition into "arts" and "science", inspite of the fact that at the N-level (between Departments) there are Joint Degree Schemes which apparently bridge the gap. At this (N+1)-level it is not an exaggeration to say that there are, and always have been two universities. As we look ahead to the degree schemes offered for 1974/5 we see, in KX(Y), that $Q_1 = 2$ also. This means that the disconnection, the split into two, <u>is going to get worse</u>. Increasing the numbers of possible degree schemes to

43 has certainly not improved the academic unity of the university.

Within any one of the two complexes (which together account for the value of $Q_0 = 2$) it is clear that the obstruction vector is practically zero. This means that it must be (and always has been) easy for students to move across the inter-School boundaries in pursuit of a degree - which corresponds to the experience of the last 4 years. But the obstacle between the School Comp-St, and the rest is very high. These facts illustrate obstruction to $\Delta\pi$, where π is a pattern (for example, of numbers of students engaged on each degree scheme) on the complexes KX(Y) or KY(X).

It is inevitable, that if $Q_0 = 2$ the members of the University, both staff and students, who work in the separate components will experience this high obstruction between them. Each will be tempted to speak of the "other side" of the University, and the resulting polarization will be a reflection of this global disconnection in the backcloth.

The impact of the School (as opposed to that of the Department) on the students is largely felt during "admission" and the first year studies. This is because the initial academic philosophy for the University was centred on the notion of a Common First Year course of study, with inter-disciplinary courses across the Departments. It is paradoxically apparent that this philosophy, aiming at one ideal of academic unity, has resulted in the de facto division of the academic community. The pressure to destroy this ideal of unity has clearly come from the Departments, as they have pursued their relatively narrow academic ambitions (no doubt with the best intentions). But the structure could have been monitored (if this structural language had existed 5 years ago) so as to have forewarned.

This kind of thing is another illustration of the <u>hierarchical</u>

implications of change. The philosophy of the "ideal of unity" is a notion appropriate to the (N+3)-level, but it must be implemented at all the lower levels to be effective. The successive "defences" to this philosophy therefore occur in the following schema:

aim of academic unity	↔	(N+3)-level concept
↓		↑
Senate		(N+2)-level decisions
↓		↑
Schools		(N+1)-level decisions
↓		↑
Departments		N-level decisions
↓		↑
Sub-Depts (Research/teaching)		(N-1)-level decisions
↓		↑
Individuals (Staff)		(N-2)-level appointments

Speaking in another context we can say that "high level managerial policy" must in the end be "implemented on the factory floor", through a well-defined hierarchical structure. In this instance, there has been no "defence" at the (N+2) or (N+1)-levels and this is certainly associated with the absense of a formal connection between (N+1) and N-levels. This absence will be discussed in more detail in other sections (v. also section 0.0) but here we can see that if the Departments were formally related to the Schools (in the sense that the Departmental Meetings were Committee Meetings with formal submissions to the School Boards) then certain kinds of decisions could not be taken at the Departmental level without there being a public discussion of the (N+3)-level policy. For example, pressure for a new kind of degree scheme or an alteration in the lecture content of an existing scheme.

For their part, the Schools have exerted their formal influence by adopting new degree schemes which were "joint schemes" between different Departments - no doubt in the instinctive hope that this is "keeping things together (connected)" and thereby giving general support to the (N+3)-policy of "unity". But without a suitable analysis of the details of any such decisions it is clear from the above tables that the struggle can be easily mistaken and the consequence unforeseen.

The consequence of taking a view of the academic studies strictly from a Departmental angle, thereby ignoring the Schools, are manifest by an analysis of the data files DP v DS (Departments versus Degree Schemes) v. pages 226/7. The following tables shows the global structure vectors for the years 1970/1 to 1974/5.

Data file: DP v DS (Y = DP = Departments, X = DS = Degree Schemes)

Year	$\underline{Q}$ for KY(X)	$\underline{Q}$ for KX(Y)	Number of DP and DS	
1970/1	$\{\overset{4}{1}\ 1\ 6\ 6\ \overset{0}{6}\}$	$\{\overset{1}{5}\ \overset{0}{6}\}$	11	20
1971/2	$\{\overset{4}{3}\ 4\ 6\ 5\ \overset{0}{6}\}$	$\{\overset{1}{6}\ \overset{0}{6}\}$	11	23
1972/3	$\{\overset{5}{2}\ 3\ 4\ 8\ 8\ \overset{0}{5}\}$	$\{\overset{1}{9}\ \overset{0}{5}\}$	12	28
1973/4	$\{\overset{7}{1}\ 2\ 3\ 3\ 6\ 9\ 9\ \overset{0}{4}\}$	$\{\overset{1}{12}\ \overset{0}{4}\}$	12	33
1974/5	$\{\overset{9}{1}\ 3\ 3\ 3\ 3\ \overset{4}{7}\ 7\ 10\ 9\ \overset{0}{3}\}$	$\{\overset{2}{17}\ \overset{0}{3}\}$	14	43

At this N-level the picture is vastly different. The most striking feature of the global structure is the relatively large values of Q_q, with the implication of large obstruction vectors to pattern changes. The values of Q_o (which are the same for KY(X) as for KX(Y) in any one year) give a new view of the University as a disconnected set of Departments (and/or of Degree Schemes). Thus there were 6 distinct 0-connected components in both 1970/1 and 1971/2, and this number has dropped steadily to 3 (for the projected year 1974/5). There is no doubt that this indicates a growing tendency to an academic unity (qua (N+3)-policy?) achieved by introducing more Joint-Degrees. But a price has had to be paid for this and (v. later in this section) one such price is an increase in the maximum value of Q_q in $\underline{Q}$. This greatest-value component has risen through the sequence

$6(q = 2)$, $6(q = 2)$, $8(q = 2)$, $9(q = 2)$, $10(q = 2)$

This means that the obstruction vector has increased dramatically at the $q = 2$ level, taking the successive values

$$0, \ 0, \ 3, \ 5, \ 7$$

What is this an obstruction to? It is an obstacle to any change in a pattern π^2 on the complexes KY(X). A typical pattern π^2 might be the "number of students per Department who were eligible to opt for 3 distinct Degree Schemes". The Departments which fall into the same 2-component are those which share 3 joint-scheme degrees. Thus, increasing the numbers of degrees from 20 to 43 has been accomplished by increasing the numbers of joint-scheme degrees, which has reduced the value of Q_0 but increased the values of Q_1 and Q_2. The disconnection has therefore been concealed, in effect, by this course of events.

The π^2 suggested above would indicate the success of the Common First Year (which is School-based and which prepares students for opting for a particular Degree Scheme for their 2nd and 3rd years) if that results in a geometry which permits changes $\Delta\pi^2$. But this structure shows a steadily increasing obstruction to both π^2 and the corresponding π^1.

The same effect is mirrored in the structure vectors for KX(Y), where the obstructions for $q = 1$ take values

$$0, \ 0, \ 4, \ 8, \ 15$$

for each year. At this q-value the typical π^1 (on KX(Y)) might be the "number of students on each Degree Scheme with serious interests in 2 Departments". Changes $\Delta\pi^1$ are increasingly obstructed by this structure. Such a student would experience a sense of being narrowly confined into one of the 12 components (in 1973/4).

The overall picture, from the global analysis, is of Departments creating a relatively narrow space for their interests via selected joint-schemes which link them (but not too much) with some others. This tendency will be manifest lower down the hierarchy by a tendency to appoint more specialist staff in existing areas of interest - a tendency which will be reflected in advertisements for and discussions about future appointments. Particular cases of this increasing rigidity are to be found in recent School Board meetings (1972/3) when the "field" for a new professorial appointment is discussed. In all cases (4 in number) the stress on existing areas (Department-wise) has been pronounced.

But if this overall interpretation is correct we would expect it to be reflected in some way in the local structure of the relation. To examine this we search for generators (q-objects/q-holes) of the shomotopy groups (S_q, *) and, in the first instance, we find the situation given in the following table.

Data file: DP v. DS q-objects (q = 0)

Year	Number of DS	Complex	Number of 0-objects
1970/1	20	KX(Y)	0
1971/2	23	KX(Y)	2
1972/3	28	KX(Y)	2
1973/4	33	KX(Y)	5
1974/5	43	KX(Y)	12

Each 0-object acts as a traffic generator and the traffic can be any of the following kinds of things:

administrative activity between Departments

academic discussions re syllabuses

academic advising of students re choice

timetabling of relevant lecture courses

use of physical rooms/labs relevant to courses
student discussion of merits of Degree Schemes
uncertainties of student choice of options
discussion of Departmental allocation of staff resources to Schemes
choice discussions re admissions of students

The actual 0-objects can be identified in the Shomotopy list on pages 201/2. It suffices to say that the simplest of assumptions would lead us to expect that between the academic years 1971/2 and 1974/5 the traffic generated in the university backcloth S(N) by the introduction of new degree schemes has increased by a factor of 6. This is largely administrative traffic which must inevitably be viewed as noise on the academic life of the community.

For the administrative staff who must handle this noise the increase in the possible number of Degree Schemes is an ominous burden which must leave them with a sense of bewilderment. Do the academic staff know what they are about, in this traffic/noise sense, when they press for more such schemes? For their part the academic staff must feel a sense of bewilderment in their efforts to achieve some (vague?) (N+3)-policy (for example, attracting more undergraduate applications) by these tactics. The reason lies in the subtle interconnections which exist in the community (and which the Q-analysis in multi-dimensional space is designed to explore in a rational way) and which so easily confound superficial attempts to guess at cause-and-effect by "simplifying" the issues.

B2.2 Committees and Structure

In discussing the Committees in the University we do not include Working Parties in the backcloth. Furthermore we take this backcloth in a typical recent year 1972/3 and refer to the following 28 Committees which are listed in the University Calender.

Level	Code	Full Name
(N+3)/(N+2)	COUNCIL	The Council of the University
N+1	SENATE	The Senate of the University
N+1	FINANCE	Finance Committee of Council
N+1	DEVELOPMENT	Development Committee of Council
N+1	CAREERS	Careers Advisory Service of Council
N+1	APPEAL	Appeal Committee of Council
N+1	ACAD-PLAN	Academic Planning Committee of Senate
N+1	ADMIS	Admissions Committee of Senate
N+1	BURROWS	Burrows and Co-operative Lectures Comm. of Senate
N+1	CARRERAS	Carreras Arab Lectureship Committee of Senate
N+1	DISC-APPEAL	Committees of Appeal (Disciplinary)
N+1	CHAIRS	Committee on Chairs
N+1	COMP	Committee on Computing Centre
N+1	LATIN	Committee on the Latin American Centre
N+1	DISCIP	Disciplinary Committee
N+1	ED TECH	Educational Technology Users Committee
N+1	LIBRARY	Library Committee
N+1	MEMBERSHIP	Membership (of the University) Committee
N+1	MEM-REVIEW	Membership Review Committee
N+1	BUXTON	Noel Buxton Lecture Committee
N+1	PENGUIN	Penguin Lectures Committee
N+1	STAFFING	Senate Staffing Committee
N+1	SOCIAL POL	Social Policy Committee
N+1	WIV-HOUSE	Wivenhoe House Committee
N+1	COMP-STUDS	Board of the School of Comparative Studies
N+1	MATHS-STUDS	Board of the School of Mathematical Studies
N+1	PHYS-SCI	Board of the School of Physical Sciences
N+1	SOCIAL-STUDS	Board of the School of Social Studies

By referring to the diagram and discussion in section B0.0 we see that the hierarchical levels of these formal committees need to be those which are listed. The Council of the University performs two functions, one at (N+2) and the other at (N+3). At the higher level of (N+3) the Council is a formal contact with the "outside" world and because of this its membership contains a high proportion of distinguished laymen. Another formal committee in the University which is not in the above list is that of the University Court. This is entirely an (N+3)-level committee and is a large body of lay-people with some University staff members. But the Council performs many functions and takes decisions which are part of the University per se and because of this it has a community role which is properly placed in the (N+2)-level beside that of Senate. It then follows that the committees of Senate and of Council will be at the (N+1)-level. Thus the set which contains the two words Senate, Council is a cover of the (N+1)-level set containing the names of the other committees. In this case it so happens that the cover is a partition.

It is obvious that we must then ask the question "What are the formal committees at the N-level, and possibly (N-1)-level, etc?" In fact there are some formal N-level committees which function as sub-committees of the (N+1)-level set. These are not in the list above but consist of

Landscape Sub-Committee of DEVELOPMENT
Building Sub-Gommittee (various) of DEVELOPMENT
Catering Sub-Committee of FINANCE
Various Sub-Committees of School Boards, etc.

Of these N-level committees only Landscape and Catering are apparently permanent, whilst the others are usual ephemeral.

This identification of hierarchical levels for the University's formal committees presents us with a significant situation. This significance lies in the complete disconnection between the lower levels (N-1) and (N-2) together with the slight and occasional connection with N-level. This means that "the University", as a structure of formal committees making official decisions on all matters, is not unlike a citadel surrounded by a wide moat but with controlled outlets to (N+3), via the Council and the Court, and to N-level via the ephemeral sub-committees of (N+1). It is particularly significant that the Departments, which exist at the N-level (because the School Boards must form a cover of the Department set) are not formally connected with (N+1). This is because the Departments do not constitute sub-committees of the School Boards or of any of the other (N+1) set. They are nevertheless required to hold at least one Departmental meeting per term but this meeting is only of an advisory nature to the Chairman of the Department. The Chairman in his turn is an ex officio member of the appropriate School Boards but he is not a member as one who brings the Minutes of his Departmental meeting to that Board. This replaces the Department (as a Committee) by the Chairman as a genuine one man sub-committee of the (N+1) School Board. But even here the word "genuine" is probably misplaced because at any School Board meeting the Chairmen of the relevant Departments are not expected to submit Minutes of their own meetings (with themselves) to that Board. A Chairman might therefore be regarded only as a de facto sub-committee of a School Board, but he clearly cannot be regarded as a de jure sub-committee.

The individual person in the University community, by which I mean anyone who is there to pursue an academic goal, is naturally found at the (N-2)-level. Much of his attachment to the community is in fact an identification with an appropriate piece of the backcloth S(N-2). In addition to this he must of course be attached also to S(N-1). For example, the student is operating on S(N-1) when he attends lecture courses and tutorial classes organised by some Department. When he is discussing the fine detail of his work he is doing so on S(N-2), where he meets the (N-2)-member of staff. The relation between a student and his tutor is clearly a mixture of hierarchical levels. After they exchange the time of day at (N-2), discuss the details of a mathematical theorem at N-2, discuss the shape and form of a lecture course at N-1, discuss departmental rules and regulations at N, discuss the ideas behind specific degree schemes, and/or facilities in the Library, etc. at N+1, discuss the grand plan of University developments and general educational matters at N+2, they finally discuss the students prospects for livelihood and happiness at N+3. When the student asks his Professor what his prospects are in society with this or that degree qualification, he is requiring guidance as to what happens when he crosses the imaginary drawbridge from the University citadel out into N+3.

It is patently clear therefore that each individual in the community is very much associated with each of the available hierarchical levels. The formal committee structure of the University is that piece of the backcloth S(N+1) and S(N+2) where decisions are taken which affect the individual at N, N-1 and N-2 levels. It is not surprising therefore that he feels occasionally the desire to participate to some extent or other in this process of planning and deciding.

Because of these general considerations we examine a particular data file for the year 1972/3 which represents the relation between a set of People and the above set of 28 committees. This set of People is the set of all the individuals who sat on any of the committees during that year and the set in fact contained 165 people. They were numbered 1 to 165 and given a code which allowed us to identify some other feature of their status such as whether they were Staff or Student member of this or that Department etc.. The data file is referred to as P v C and the identities of the members are irrelevant except in so far as we wish to refer to the office which anyone might in fact hold. Thus I hope that we have shown proper respect by denoting the office of Vice-Chancellor by number 1 and other than that identification of officers is only incidental to the general discussion. The complete Q-analysis of this data file is listed on pages 201/2 and the Q-objects which were found are listed on pages 230/2 .

<u>Data file</u>: P v C (1972/3) $KY(X) \equiv K_P(C)$

The expected and obvious fact which emerges is that the Vice-Chancellor is the simplex of highest dimension in the structure with $q = 19$. This polyhedra with 20 vertices dominates the structure until we reach the value of $q = 9$. Then we see that two distinct components are added by persons 22H and 28L. These people were respectively members of the History and Literature Departments. The structure which unfolds gives us the global structure vector:

$$Q = \{\overset{19}{1} \ldots \overset{10}{1} \;\; 3 \;\; 2 \;\; 3 \;\; 3 \;\; 6 \;\; 4 \;\; 5 \;\; 9 \;\; 7 \;\; \overset{0}{1}\}$$

This gives us an obstruction vector of

$$\underline{\hat{Q}} = \{\overset{9}{2} \;\; 1 \;\; 2 \;\; 2 \;\; 5 \;\; 3 \;\; 4 \;\; 8 \;\; 6 \;\; \overset{0}{0}\}$$

What does this vector $\hat{Q}$ actually obstruct? It is the obstruction to changes in any pattern π which is defined on the people and is a function of the vertices, which in this case are the committees. Such a pattern might represent particular items of business which are considered by a selection of committees and which individual people give a ranking to. The values of the pattern π on the individual people might then be these ranking values which they attribute to the items of business. Change for π, $\Delta\pi$, therefore amounts to a change in the ranking which individual people are attributing to items of business which the committees are considering. A high obstruction vector means that the system is that much more rigid and that a change of such rankings is more difficult to achieve. This clearly represents the apparent <u>inertia of policy</u> with regard to a set of issues. Since we do not attribute a single value to an obstruction vector we interpret this situation by referring to the individual component values of $\hat{Q}$. We notice here that the largest such component occurs at $q = 2$. This refers to items of business which are considered by three committees. The next highest value occurs at $q = 1$ and this refers to items of business which need to be considered by two committees. The whole of the obstruction vector $\hat{Q}$ shows a reasonably high level of obstruction at any q-value and its length is an indication of how far the rigidity, or inertia, exists throughout this structure. The maximum q-value of 9 refers to items of business which need to be considered by 10 of the committees. Since this kind of item is likely to be unusual it is fair to deduce that the obstruction to a change of ranking pattern is non-zero and reasonably high for all items of business. Changing the pattern π means that individual people need to alter their ranking of a particular item of business, they need to change their minds, and this structure vector shows that it was unlikely to happen in most cases.

What we are describing is a <u>structural</u> feature derived from the data file. We are not describing the rigidities which might be built into an individual person at some (N-2)-level. The individual people who sit on these committees and thereby function at the (N+1)-level are themselves subject to a sense of being caught or trapped in the "system". This system is the structure which we are here listing and it is a multi-dimensional constraint on the way in which these people have to function and arrive at decisions. As they struggle to arrive at decisions over a set of specific issues the rigidities of the system, indicated by the obstruction vector $\hat{Q}$, impose themselves on the decision-making process and are made manifest by the individual people feeling the constraints as a tendency to rigidity, conservation and apparently relative helplessness. Without wishing to devalue in any way the responsibility of reaching decisions we can nevertheless sum up that process as one which requires the readjustment of ranking patterns π such as we have already described. Speaking somewhat loosely it is the kind of sophisticated "horse trading" between priorities. A high obstruction component is equivalent to a severe restriction on this trade-off and it is not surprising therefore that decisions are commonly arrived at by appealing to "what is usually done". This is another way of expressing the high obstruction.

<u>Data file</u>: P v C 1972/3 $KX(Y) \equiv K_C(P)$

The highest dimensional simplex in this complex is that of Senate with q = 36 and the next one occurs at q = 30 with Chairs. The global structure vector is as follows

$$\underline{Q} = \{\overset{36}{1} \ldots \overset{30}{2} \ldots \overset{27}{3}\ 4 \ldots \overset{18}{6}\ 6\ 8\ 8\ 10\ \overset{10}{12}\ 9\ 13\ 13\ 11\ 8\ 5\ 4\ 7\ 2\ \overset{0}{1}\}$$

The corresponding obstruction vector is

$$\hat{\underline{Q}} = \{\overset{30}{1} \ldots \overset{27}{2}\ 3 \ldots \overset{18}{5}\ 5\ 7\ 7\ 9 \ldots \overset{10}{11}\ 8\ 12\ 12\ 10\ 7\ 4\ 3\ 6\ 1\ \overset{0}{0}\}$$

As in the conjugate complex this shows relatively high obstruction at most of the q-values, notably at q = 6, 7, 8. Two committees share the q-face if they share (q+1) people out of the set of people. If there is a large number of components at some q-value it means that the structure KX(Y) falls into that number of separate components and a committee out of one component cannot share (q+1) people with any committees out of any other component. The obstruction which is therefore manifest must therefore refer to matters which are functions of the committee members (the people) and are required a given number of members before they can be transmitted from one committee to another. This is a generalisation of the idea of calling individual votes to settle an issue as it marches through the committee structure. Voting for some measure on a committee is an illustration of the rearrangement of rankings which the individual performs when he is required to make a decision in the matter. We have seen that his ability to do this is closely constrained by conjugate structure KY(X). In this structure we are now considering, KX(Y), the formal business conducted in the committees and the formal decisions which are taken on that business correspond to patterns (in suitable ranking valuation) on the committees - that is to say on the simplices of the complex KX(Y). Hence the rigidity in the system, which is indicated by the component values of the obstruction vector, are rigidities associated with getting certain <u>kinds</u> of business through the structure. This is conjugate to the kind of rigidity discussed in KY(X).

We are here seeing the effect of having particular individuals on the various committees. The global structures we are looking at represent the multi-dimensional connectivities which are thereby set up. It is in this multi-dimensional geometry where the decisions occur, where the "action" takes place. It follows that as we contemplate the movement of people about the committees from one year to the next we are in fact witnessing a distortion of the geometrical structure so far described. This distortion will in the global sense be described by the structure vectors such as those above and the obstruction vectors will indicate changing kinds of rigidity or inertia in the system. This means that when each summer term comes around and the proposed list of committee members is publicised a careful monitoring by our Q-analysis would give us an official indication of how the machinery would be constituted and how it would function during the next academic year. Surely in this way we would be able to plan for and achieve any desired level of rigidity/flexibility in the system. If we wish to have a completely rigid system, one which from the decision-making point of view was ultra-conservative, then we would be able to achieve it by adjusting the obstruction vector to be as large as possible. On the other hand if we wish to have a system which was extremely flexible and capable of rearranging its priorities under a pressure of events then we would ensure as low an obstruction vector as possible. These kind of considerations would give us constraints on the structure which were necessary rather than sufficient.

The high obstruction value at $q = 7$, in KX(Y), namely the value of 12 means that the pattern

$$\pi : \text{COMMITTEES} \rightarrow \text{VALUES}$$

which was also a function of people would be most obstructed at this level. This is the level at which committees share 8 members in pairs. Hence if π denotes a pattern on an item of business which depends upon 8 people it is going to be relatively difficult for its values to change in the global sense. This is like saying that the prospect of feeding through the issue which requires 8 votes for success is in a relatively rigid structure. That is to say that if it can only be achieved by a rearrangement of existing rankings by individual persons then the structure which is inherent offers a considerable obstacle to its success. This is the reason why the canvassing for some item of business needs to be done outside the formal committee structure, in the Coffee bars and licenced premises. Presumably these latter places act as informal committees with a structure which offers a low obstruction vector. The more obstruction there is in the formal structure the more items of business need to be carried outside and discussed in a more flexible situation. This results not unnaturally, in a rigid system, in the feeling that all the political decisions are being taken informally in other places by unnamed individuals and that when the committees meet the business appears to have been already settled. One consequence of this will clearly be that members of committees must be spending at least twice as much time and energy on considering specific items of business as they need to do. It is therefore difficult to draw any conclusion other than the desirability of a low obstruction vector in these particular kinds of structure.

....

We can also obtain some information from a consideration of the Eccentricities of the various simplices in both KY(X) and KX(Y). If a simplex possesses a top $\hat{q}$ and a bottom $\check{q}$ then we recall that its eccentricity is given by

$$\mathrm{Ecc}(\sigma) = \frac{\hat{q} - \check{q}}{\check{q} + q}$$

and this is a measure of the extent to which the simplex σ is "out of step" with its fellows or, from the other side of the fence, how well it is integrated into that local part of the structure to which it is connected.

In KY(X) we find that the vast majority of the individuals have 0-eccentricity and we merely list the first few individuals who have the highest top-q values. This gives us the following values

Ecc	Individual
$^{3}/2$	1
$^{1}/9$	22H
$^{2}/7$	28L
$^{1}/7$	23M

It is perhaps not surprising that the Vice-Chancellor possesses the largest eccentricity in this structure. His top q-value is 19 whilst his bottom q-value is 8. This is the largest bottom q-value for any individual so that it is fair to say that he is more highly connected to his fellows in this structure than is anyone else. On the other hand his top q-value shows him to be more remote from his fellows than is anyone else, vis-a-vis the formal committees.

For the structure KX(Y) we list the major committees in descending order of their eccentricities

Ecc	Committees
14/5	SOCIAL-POL
12/5	DISCIP
2	COMP STUDS
5/3	PHYS-SCI
4/3	COUNCIL
21/16	SENATE
15/16	CHAIRS
9/10	MATHS-STUDS
7/10	SOCIAL-STUDS
2/7	FINANCE
2/10	DEVELOPMENT
1/8	STAFFING
0	ACAD-PLAN

In looking at this list we must remember that what committees share is committee members. It follows that a high eccentricity means that relative to its fellows a committee contains a high number of people which are not to be found on any other committee. In comparison a low eccentricity means that the members of the committee are to be found by-and-large on other committees. The smallest possible eccentricity is zero and this is achieved by ACAD-PLAN whilst the highest is that of SOCIAL-POL with a value of 14/5. If we concentrate on the (N+1)-level committees, we see that the bottom end of the list corresponds to those committees which people tend to regard as influential whilst the top end indicates those committees which seem to produce the most noise (discussion of business items) and probably the least results (in the sense of having recommendations accepted at the higher (N+2)-level).

....

We can also take the data file P v C and deduce from it another file C v DP of the 28 committees versus 16 Departments - using the word Departments in a broad sense. This is simply done by noticing which Department any individual person is associated with and thereby obtaining a matrix which has the number of people per Department per Committee as its entries. We use the following notation

Y = Committees (28)   X = Departments (16)

where set X is :	AD	(Administration)
	MA	(Mathematics)
	GOV	(Government)
	ECON	(Economics)
	SOC	(Sociology)
	CC	(Computing Centre)
	LIT	(Literature)
	CH	(Chemistry)
	LA	(Language Centre)
	EE	(Electrical Engineering)
	HIST	(History)
	PH	(Physics)

LAY	(Lay people)
AR	(Art)
ST	(Students)
LIB	(Library)

The Q-analysis requires a slicing parameter since the data file corresponds to a matrix with entries not restricted to 0/1. This parameter is denoted by SV (Slicing Vector) and, for example, if SV = 2 then the corresponding structure says that a Department is only a member of a Committee if it has at least 2 members on that Committee.

We obtain the global pictures as follows (the Q-analyses are shown on page 233/5).

<u>Data file</u>: Y v X (SV = 1)

KY(X) $\underline{Q} = \{\overset{12}{3}\ \ 2\ \ 1\ \ 1\ \ 4\ \ \overset{7}{5}\ \ 1\ \ldots\ \overset{0}{1}\}$

KX(Y) $\underline{Q} = \{1\ \ldots\ \overset{15}{3}\ \ 5\ \ 8\ \ 7\ \ 6\ \ \overset{10}{4}\ \ 2\ \ 2\ \ \overset{7}{1}\ \ldots\ \overset{0}{1}\}$

<u>Data file</u>: Y v X (SV = 3)

KY(X) $\underline{Q} = \{\overset{7}{1}\ \ 1\ \ 2\ \ 2\ \ 2\ \ 3\ \ 3\ \ \overset{0}{1}\}$

KX(Y) $\underline{Q} = \{\overset{5}{2}\ \ 4\ \ 5\ \ 5\ \ 3\ \ \overset{0}{1}\}$

These structures correspond to the view that individuals go to a Committee meeting "wearing a Departmental Hat"; that they go as N-level entities to discuss N-level business - and thereby define the (N+1)-level Committee as a structure in S(N). When we take SV = 3 we are effectively saying that the "Departmental view" does not count until there are at least 3 individuals from the Department on that Committee.

In the light of our discussion above we can summarise the consequences in the following way:

(1) For SV = 1 there is zero obstruction to all $\Delta\pi^t$ up to t = 6 in both KY(X) and KX(Y).

(2) Trade-off between Departments on business items is not obstructed until q = 7.

(3) This depends upon all Departmental representatives speaking with one voice.

(4) At SV = 3 the Departments with more than 2 representatives on a Committee do all the "horse-trading".

(5) The obstruction to changes in any $\Delta\pi$ is larger than in the case of SV = 1 for <u>all</u> values of q.

(6) Do we conclude that more flexibility is built into the system by restricting Departmental representation to 1?

(7) This supports the notion that flexibility is improved by making the Departments formal sub-committees of the (N+1) set.

(8) Overall these structures show that the "Departments" as a set are more influential than "Individuals" as a set. This is because of the practically zero obstruction vectors,

$$KY(X) \quad SV = 1 \quad \hat{\underline{Q}} = \{\overset{12}{2}\ 1\ 0\ 0\ 3\ 4\ 0\ 0\ 0\ 0\ 0\ 0\ \overset{0}{0}\}$$

$$KX(Y) \quad SV = 1 \quad \hat{\underline{Q}} = \{\overset{15}{2}\ 4\ 7\ 6\ 5\ 3\ 1\ 1\ 0\ 0\ 0\ 0\ 0\ 0\ 0\ \overset{0}{0}\}$$

B2.3 Committees and q-objects

The shomotopy analysis of the data file C v P (1972/3) is summarised on pages 201/2. It shows that, during that academic year, there were 10 q-objects in $K_C(P)$. In the derived data file C v DP (Committees versus Departments) there were 2 0-objects at SV = 3 and 1 0-object at SV = 1, all in $K_C(DP)$. Each of these objects is a "hole" in the structure and we know from our discussion so far that it acts as a <u>traffic generator</u>, in the local sense.

As an example we look at the 4-object found in $K_C(P)$ with generator

G1 : {AC/PL CHAIRS MA/ST CPTG SENATE AC/PL}

These committees form a 4-loop ($q = 4$) in $K_C(P)$ and this means, for example that

AC/PL and CHAIRS share a 4-face

Hence there are 5 people (from the set P) who simultaneously sit on both of these committees. Similarly there (at least) are 5 people common to the other neighbouring pairs. But the significance of the loop (or "hole") is that there is no committee which fills the hole (at this q-level). This is shown diagrammatically in Figure 10, where the presence of the q-hole is represented by the shading.

In the (abstract) space defined by $K_C(P)$ the hole (at $q = 4$) provides a boundary to certain relevant action. Thus an item of business which interests (not less than) 5 people on two committees on opposite sides of the loop (for example, AC/PL and MA/ST) cannot find a single home (committee) in which it could be discussed - the local geometry does not offer a place for it. That item of business, in that space, cannot have a geometrical identity but must therefore exist in (at least) 2 places. The class of such items of business can therefore only find that "place" in the geometry which includes the whole of the loop. Thus this class of business automatically becomes "traffic" which goes round the q-objects.

Once more we can illustrate the appropriateness of using the word "object" as synonymous with "hole". This class of business must "bounce off" the hole because it cannot go through it, the hole is opaque, it feels like an object of everyday experience. Thus the word "q-hole" is a proper description of a property of the static geometrical backcloth, whilst the word "q-object" properly describes the experience of that hole by dynamic patterns on that geometry.

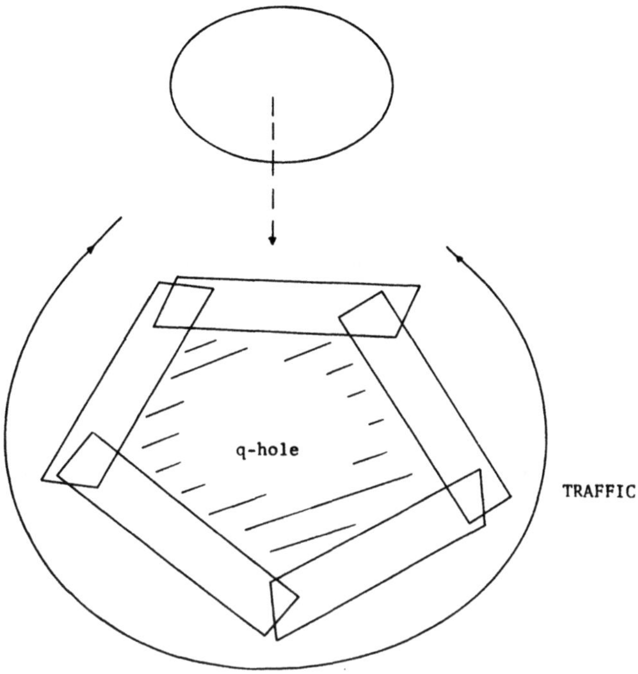

Figure 10. Shomotopy Hole: q-object and pseudo-committee

This would suggest that the higher q-value-objects in $K_C(P)$ are relevant to traffic which is more significant in the community, because such traffic is made up of items of business (of community issues) which are of concern to more people. The filling of a q-hole (if that is possible) is likely to be of more value (and to be appreciated by more people) for the larger q-values than for the smaller q-values. Since the q-objects are associated with a certain hierarchical level ((N+1) in the instance cited) it is clear that any (N+2)-item of business must fall into the category of those things which help to make up the traffic round those objects. It therefore becomes more difficult for the Committee which make up a q-hole to reach a decision on any such (N+2)-item without help from some outside source. In this instance, typical (N+2)-business would be quinquennial plans, making new fields of research, dealing with the (N+3)-image of the university in Society at-large, financing the university, etc..

But items of business need to be not only discussed but also settled (decided upon). The practical running of the university therefore needs a steady supply of decisions on all matters, including those items which constitute traffic round the above 4-object. How is this achieved? By creating a new committee to fill the 4-hole. But the geometry of $K_C(P)$ does not contain such a committee. Therefore a "pseudo-committee" must be introduced - which is not part of the formal structure (by definition). This pseudo-committee occupies the 4-hole, sits on the 4-object, becomes the 4-object, and naturally takes charge of the traffic around it. This "taking charge of" means being the instrument of decision-making re the traffic.

But a pseudo-committee must be a set of people and each person on it should be sufficiently q-connected with the interfaces (sets of other people) between the committees in the q-loop as to make the pseudo-committee a proper hole-filler. In this particular case this means that the greatest number of people in the 5 interfaces does not exceed 25 (=5 x 5). The pseudo-committee should therefore consist of people who share faces with these 25 people - sharing those faces in a structure outside the formal $K_P(C)$.

This probably means that there is only one reliable candidate for the pseudo-committee, the Vice-Chancellor, and this argument seems to suggest that the structure of $K_C(P)$ thrusts this role upon the man who occupies that office. In this case we are led to say that, in $K_C(P)$, the Vice-Chancellor is primarily the person who occupies the q-holes - and the larger q becomes the more likely is it that he becomes the only member of the pseudo-committee. Other office-holders in the university are probably candidates to join the VC in other pseudo-committee hole-fillers, for smaller q-values.

But this situation in which, for the sake of making the point as forcible as possible, one person plays the role of various q-objects,is one which suggests strongly an interpretation of personal power or influence.

From this object's point of view (that of the pseudo-committee which occupies it) the traffic around it illustrates the difficulty which colleagues have in discussing business and of arriving at sensible and consistent decisions. It is therefore a prime responsibility of that q-object to initiate the decisions. This attracts the traffic towards the q-object and thereby brings it to a halt.

From the point of view of the people on the loop-committees (and all loops shomotopically equivalent to that one) it appears that there is a source of power, of decision-making, which is "behind the scenes" - the pseudo-committee (which cannot be formally identified). Thus the Vice-Chancellor (for example) becomes credited with this apparently sinister role of "pulling the strings" and of "getting his own way" in some mysterious way. If in fact the pseudo-committee consists of the VC alone then it is true that he will inevitably "get his own way" at $q = 4$ by being the above 4-object. The alternative is to remove the 4-hole from the backcloth by an alteration to the structure $K_C(P)$.

The only way of bringing this discussion into public view is to monitor the People v Committee Structure before the event.

If we look at the shomotopy analysis on page 201 we see too that there are 2 3-objects in $K_C(P)$ with XR = 3. This XR is a SHOM-program parameter which excludes certain of the rows in the data file P v C. In this case 3 rows (people) were excluded and these were those with code names

1	(the Vice-Chancellor)
22H	(the Dean of Comp-Studs)
28L	(the Director of Admissions)

The result was to "unearth" 2 3-holes in $K_C(P)$. Hence, at $q = 3$, this meant that the pseudo-committee consisted of

{1, 22H, 28L}

and so this set formed a "behind the scenes" 3-object. They exercised real power/influence at the $q = 3$ level around the objects defined by

G1 : {AC/PL, STAFFING, CHAIRS, CPTG, MA/ST, SENATE, AC/PL}

G2 : {AC/PL, STAFFING, CHAIRS, SOC/ST, DISCIP, SENATE, AC/PL}

It is not proper to lay "blame" on these individuals for this situation - unless we accept the unlikely hypothesis that they so understood the structure as to have consciously manipulated it to this end - but to realise that the structure thrusts this role upon them.

The fact that a pseudo-committee (such as the VC) plays a decisive role at some q-level of q-objects does not imply that it plays an equally decisive role at (q-1)-level or (q-2)-level, etc.. Indeed, the study of the q-objects in C v P shows that the 4-object is not a 3-object, nor are the 3-objects the same as the 2-objects. Thus there is a consequent interplay of q-objects at various q-levels, with a consequent mix-up of the necessary pseudo-committees. This probably results in the highest q-level q-objects being the most influential for obvious reasons.

The fact that the SHOM-program found 10 q-objects in $K_C(P)$ for 1972/3 shows that the committees in the formal structure (28 of them) must have been largely ineffective for all classes of items of business which legitimately concerned (not less than) 3 people. Can we therefore ask, "what was the practical point of the Committees?"?

For individuals in the University (not necessarily members of P) it is inevitable that they see the decision-making process as a mysterious one which functions "behind the scenes" - because in fact it functions in the 10 q-objects - and that the invitation to sit on a Committee (to join in the political life of the community) is nothing more (in fact, if not in intention) than an invitation to join the traffic round the q-objects. Thus the traffic becomes noise around the actual decision-making - which is effectively done by the q-objects.

Unfortunately it is almost certainly true that the officers of the University who issue these invitations do so with the best of intentions and with the sincere desire to encourage a "democratic participation" in the structure. They find it difficult to be sympathetic to the student/staff who turn sour on "the system" and who complains of the lack of "genuine" participation. Exhortations to such people to "make full use of their democratic opportunities on the committees" can unfortunately often mean only an exhortation to be more energetic as traffic. If the individual is anxious to make <u>decisions</u> (to exercise responsibility) then he is really anxious to be an object which fills a q-hole, and it could well be that the more he senses his role as traffic (which actually allows some other object to exist) the more frustrated is he likely to feel.

In moving from the structure of C v P to that of C v DP (Committees versus Departments) at SV = 3 we notice (v. page 202) that there are 2 generators of 0-objects

G1 : {AC/PL, CHAIRS, COMP-ST, DISCIP, SENATE, AC/PL}

G2 : {AC/PL, COUNCIL, MA-ST, DISCIP, SENATE, AC/PL}

These refer to the structure of Committees when they are seen as made up of departmental delegates - and ignoring less than 3 such delegates on any one committee. It implies that the business of the Committees is settled by departmental considerations (at the N-level).

B2.4 Committees and Business

To obtain an idea of the items of business dealt with by the university committees, we undertook the task of carefully examining the Minutes of every meeting of Senate and of Council since the university began (the years 1964-1973) and, in addition, the Minutes of the following 11 committees for the years 1970-1973,

Committee Set C :	1	DEVELOPMENT	N + 1
	2	BUILDING (Sub-Cttee of 1)	N
	3	LANDSCAPE (Sub-Cttee of 1)	N
	4	FINANCE	N + 1
	5	CATERING (Sub-Cttee of 4)	N
	6	ACAD/PLAN	N + 1
	7	SOCIAL-POL	N + 1
	8	MATHS-STUDS	N + 1
	9	PHYS-SCI	N + 1
	10	COMP-STUDS	N + 1
	11	SOC-STUDS	N + 1

The items of business considered by these committees of C fell naturally into the two levels of N and (N-1). That is because, for example, an (N+1)-level Committee exists at that structural level by virtue of its discussion being between entities at the N-level - these entities being "people wearing Departmental hats" and "items of business". We therefore obtained the two Data files,

C v B (at (N-1)-level) and C v B (at N-level)

whose vertices are listed on pages 152 and 153 . In the N-level set (page 152) the item "30 Qq-development" refers to any appropriate Quinquennial Planning Business, such as the N-level details of the proposal to increase the number of Departments in the University. Under "29 NEW DEPARTMENTS" would come discussion of specific Departments. In the (N-1)-level set of Business (page 153) "6 UGC NORMS" means the norms in building standards etc,set by the University Grants Committee, "U/G" refers to Undergraduate matters, "G" refers to Graduate matters, "WH" refers to Wivenhoe House, whilst "16 HEX" refers to the Hexagon Restaurant matters.

The Q-analysis for both Data files are listed on pages 236/239, where the Committees are referred to by their numbers (such as 1 CTTEE, 2 CTTEE, etc.) - this being the format of the computing output.

The structure vectors for C v B, at (N-1)-level, are

$$K_C(B) : \underline{Q} = \{\overset{14}{1} \quad 1 \quad 3 \quad 4 \quad 4 \quad 4 \quad 4 \quad \overset{7}{5} \quad 4 \quad 4 \quad 4 \quad 3 \quad 3 \quad 3 \quad \overset{0}{2}\}$$

$$K_B(C) : \underline{Q} = \{\overset{3}{1} \quad 3 \quad 4 \quad \overset{0}{2}\}$$

There are 2 distinct components at q = 0. In C these consist of

(i) the 4 School Boards (8, 9, 10, 11) and

(ii) the rest (without ACAD/PLAN), (1, 2, 3, 4, 5, 7)

<u>Data file : C v B</u> <u>N-level Business</u>

<u>Rows = 11</u> <u>Cols = 42</u>

1	DEVELOPMENT	1	U/G NUMBERS	31	DEGREE COURSES
2	BUILDING	2	G NUMBERS	32	EDUCATION STUDIES
3	LANDSCAPE	3	U/G WITHDRAWALS	33	ENVIRONMENTAL STUDIES
4	FINANCE	4	SCHOOL QUOTAS	34	PROGRESS CTTEE
5	CATERING	5	ADMISSION/RECRUITING	35	EXAMINERS
6	ACAD/PLAN	6	ACADEMIC DEPARTMENTS	36	CONFERENCE
7	SOCIAL POLICY	7	ED-TECH-UNIT	37	CFY
8	MATHS/ST	8	LATIN AMERICAN CENTRE	38	JUN ACAD STAFF
9	PHYS/SCI	9	MATERIAL SCIENCE UNIT	39	PARKLAND
10	COMP/ST	10	DIP HE	40	SERVICES
11	SOC/ST	11	ADMINISTRATIVE STAFF	41	MA COURSES
		12	PENSIONS	42	UNION BUILDING
		13	SALARIES		
		14	ALLOWANCES		
		15	WAGES		
		16	LOCAL INVESTMENT		
		17	CITY INVESTMENT		
		18	LOAN-FINANCE		
		19	NEW-ACCOM		
		20	HEALTH CENTRE		
		21	RESIDENTIAL-ACCOM		
		22	BUS SERVICES		
		23	HOUSING		
		24	SPORTS FACILITIES		
		25	RETAIL		
		26	CATERING		
		27	ADVISERS		
		28	COUNSELLORS		
		29	NEW DEPARTMENTS		
		30	Qq-DEVELOPMENT		

<u>Data file : C v B</u> <u>(N-1)-level</u>

<u>Rows = 11</u>

1 DEVELOPMENT
2 BUILDING
3 LANDSCAPE
4 FINANCE
5 CATERING
6 ACAD/PLAN
7 SOCIAL/POLICY
8 MATHS/ST
9 PHYS/SCI
10 COMP/ST
11 SOC/ST

1 AUDITORIUM
2 AVON WAY DESIGN
3 ACCOM-SITING
4 PROJECT TIMETABLE
5 TENDERS
6 UGC NORMS
7 FURN/FITTINGS
8 ROOM NUMBERS:
9 TOWERS
10 LAUNDERETTE
11 LITTER BINS
12 BAR GLASSES
13 DISH WASHERS
14 MENUS
15 BAR FIXTURES
16 HEX
17 WH BAR
18 WH RESTAURANT
19 BAR SQUARE 3
20 RESTAURANT SQ 3
21 COFFEE BAR
22 SITE OFFICE
23 GRASSLAND
24 FLOWERS
25 TREES
26 EXT-LIGHTS
27 PATHS
28 SEATS
29 TRAFFIC CONTROL
30 CAR PARKS
31 SCULPTURE
32 LAKES
33 EXT-DECOR
34 IRRIGATION
35 PLAY AREA
36 DAY NURSERY
37 TEACHING TIMETABLE
38 EXAMINATIONS
39 VACATION GRANTS
40 COURSE SUBJECTS
41 ASSESSMENT RULES
42 TOWERS RENTS
43 INTERVIEWS APPLICANTS
44 EXAMINERS
45 U/G INDIVIDUALS
46 G INDIVIDUALS
47 CFY SUBJECTS
48 A-LEVEL ENTRY
49 RESIT EXAMS

The ACAD/PLAN committee discussed no business at (N-1)-level, and the School Boards discussed no business which was discussed by any other committee. This reflects the natural break-down of Business into academic and non-academic issues, at (N-1). This breakdown becomes itemised under $K_B(C)$ at $q = 0$ where one 0-component contains the items numbered

37, 38, 39, 40, 41, 43, 44, 45, 46, 47, 48, 49

This means that the discussion of Business split the university into 2 disjoint pieces, each piece forming a channel for certain (N-1)-business - a channel which leads to Senate/Council at (N+2). At the level of $q = 1$ in $K_B(C)$ the components identify the pairs of committee which provide the channel (in the abstract geometry) as follows:

(N-1)-business	Committees in C	(N+2)-goal
{2, 3, 4, 5, 6, 7, 8}	BUILDING DEVELOPMENT	COUNCIL
{9, 12, 13, 14, 15, 16, 17, 18, 19, 20, 21}	CATERING FINANCE	COUNCIL
{22, 30, 33, 36}	LANDSCAPE DEVELOPMENT	COUNCIL
{37, 38, 39, 40, 41, 43, 44, 45, 46, 47, 48, 49}	SCHOOL BOARDS	SENATE

The structure vectors for C v B, at N-level, are

$$K_C(B) : \underline{Q} = \{\overset{14}{2}\ 2\ 3\ 1\ 2\ 4\ \overset{8}{3}\ 3\ 3\ 3\ 3\ 4\ 4\ 2\ \overset{0}{1}\}$$

$$K_B(C) : \underline{Q} = \{\overset{5}{1}\ 1\ 3\ 3\ 2\ \overset{0}{1}\}$$

At this level the complexes form only 1 component at $q = 0$. But at $q = 1$

there are 2 components in $K_C(B)$, one of which consists of the union of ACAD/PLAN with the four School Boards. In $K_B(C)$, at q = 1, one component consists of the set of business items numbered

19, 20, 21, 22, 26

It follows that there is one connected university structure for the dimensions of N-level Business, compared with two for (N-1)-business.

But the overall picture can be summarised by saying that the (N-1)-level and N-level Business moves on the Committee Structure via the complexes $K_B(C)$ at all levels up to q = 5. These business items constitute the traffic on the structures

$$K_C(P),\quad K_P(C),\quad K_C(DP),\quad K_{DP}(C)$$

as well as in the structures

$$K_{DS}(DP),\quad K_{CP}(DS)$$

already discussed in previous sections.

Because of this, as we would expect, Shomotopy analysis shows that there are no q-objects in $K_C(B)$ or $K_B(C)$, either at (N-1) or at N-level.

But these items of business in the sets B (at (N-1) and N) move in the geometrical spaces which enfold the q-holes already discussed in $K_C(P)$ etc..

B2.5 Local q-objects in administration

Although it would seem that the function of the administration is to provide relevant traffic (or to keep the channels of communication open - flowing?) in the spaces S(N) etc., of the Committee Structure, that function is not the whole story. We have seen in earlier sections that, in a manner which we cannot claim to fully understand as yet, the chief generators of traffic are the q-holes (via shomotopy) in S(N). Such q-holes, in addition, create opportunities (or impose the need) for decision-making by pseudo-entities which occupy those holes (informally).

In such circumstances the individuals who occupy the q-holes feel it as the burden of responsibility, whilst their opponents are bound to see it as the exercise of personal (or group) power.

Not surprisingly therefore we find the occurrence of q-holes within the administrative service itself. Their existence would suggest

(i) they generate traffic in the traffic (is that bound to be noise?)

(ii) they force responsibility/power onto individuals.

As an illustration, we take a Data file, AD v AD, in which the various administrative officers form the set AD. Then any member $X \in AD$ is related to another member $Y \in AD$ if, and only if, X and Y consult with each other in the course of their work. The data required for this Data file was obtained by a sequence of interviews with administrative personnel.

The relation is really between the <u>offices</u> (rather than the officers) and the following q-objects are self-explanatory.

At q = 7 G1 : {7 9 12 21 7}

At q = 5 G1 : {5 6 7 8 5}

G2 : {6 7 15 13 14 6}

with the key:

5 : academic registry, office dealing with senior staff and Council,

6 : academic registry, office dealing with examinations,

7 : office dealing with printing centre, administration of Disciplinary Committee,

8 : central office of admissions,

9 : Dean of Students Office, welfare

12 : office in charge of catering,

13 : office for administration of residential towers,

14 : office dealing with junior staff (salaries and wages),

15 : office of (deputy) finance,

21 : office dealing with maintenance and repairs.

To discover which set of people in the administration occupy these q-holes we add in a selection of possibilities to see if the addition results in the hole disappearing. This resulted in the following list:

Registrar
Finance Officer
Estates & Planning Officer
Academic Registrar

These four officers in the administration, or various subsets of them, must act de facto as the 5-objects and 7-objects described above.

B3.0 Patterns of Finance on the Hierarchical Structure

If we interpret the income/expenditure accounts of the university's finances as dynamic patterns

$$\pi : S(N) \to £J$$

J being the integers, then we see that π is both graded (partitioned on S(N)) and filtered (over all N-levels). The Annual Accounts, presented to the University Court at (N+3), are represented via traditional accounting methods as a mixture of both grading and filtering over all N-levels and (unspecified) q-values.

The following tables show this scheme of things in terms of our hierarchical levels.

Level	Expenditure Pattern π_m
N + 3	$\pi_{N+3} = \pi^o + \ldots + \pi^t$
	$\cup$
N + 2	$\pi_{N+2} = \pi^o + \ldots + \pi^s$
	$\cup$
N + 1	$\pi_{N+1} = \pi^o + \ldots + \pi^k$
	$\cup$
N	$\pi_N = \pi^o + \ldots + \pi^p$
	$\cup$
N - 1	$\pi_{N-1} = \pi^o + \ldots + \pi^q$
	$\cup$
N - 2	$\pi_{N-2} = \pi^o + \ldots + \pi^h$

↑ Filtration ↓

← Grading →

Income	Level	Expenditure
University Grants Cttee	N + 3	Investments in City Stock
Local Authority Grants	N + 3	Investment in Property
Private Covenants	N + 3	
Investment Profits	N + 3	
NIL	N + 2	Senate/Council, Administration Staff Salaries, Hospitality
NIL	N + 1	Schools Administration, New Buildings, Development, Landscape
Research Grants	N	Departmental Recurrent Grants
Local Rates Refund	N	Departmental Research Expenses
	N	Staff Salaries, Social Amenity Admin.
	N	Maintenance, Power supplies
Accommodation Receipts	N - 1	Foods, Drinks (Catering), Salaries
Catering Receipts	N - 1	Materials (Maintenance)
Individual Rents	N - 2	Prizes, Lectures, Studentships
Personal Payments (Catering)	N - 2	Individual Staff Salaries
Student Fees	N - 2	Itemised materials, catering

What is commonly referred to as "sharing out the cake" is an exercise which takes place at a particular hierarchical level. For example, at the N-level the Vice-Chancellor discusses with the Chairmen of Departments their expenditure requirements for the coming academic year. At this level the cake-sharing corresponds to finding a partition (a grading) of the pattern π_N. But this kind of partitioning is also subordinate to the filtration inherent in these patterns π_r on the hierarchical r-levels. Thus the total expenditure is a π_{N+3} with grading on the structure S(N+3) - vertices taken out of the items in the first table above. This expenditure π_{N+3} must include all of the expenditure at (N+2)-level, π_{N+2}, etc.. Eventually all the expenditure will be accounted for in the pattern π_{N-2}, on S(N-2). Speaking loosely, the expenditure sums are filtered down the hierarchy some being trapped at each level.

In the published audited accounts the filtration which is effectively used encompasses only the scale

$$N + 3 \quad N + 2 \quad N + 1 \quad N$$

and the grading varies from one heading to another, as one would expect. Thus, for example, under Catering (at N-level) there are listed 6 vertices, viz., food, beverages, beer, wines, spirits, tobacco, and these are grouped (into simplices)

$$\langle \text{food, beverages} \rangle, \langle \text{beer, wines, spirits, tobacco} \rangle.$$

The expenditure is then given as a pattern with the grading

$$\pi^1 + \pi^3$$

corresponding to these simplices.

It is therefore possible to read the Accounts by reference to the filtration/grading and thereby to obtain an idea of the t-forces which are reflected in changes $\Delta\pi$ in this particular pattern of expenditure. Such t-forces are the real consequences of the decisions which are taken to "slice the cake", remembering too that any t-force at the N-level must be equivalent to a p-force at the (N-1)-level, and that $p \geq t$.

For example, if we compare the expenditure on Catering (at N-level) between the two years 1971/2 and 1972/3 we see that there was a 3% fall on the above $\pi^1 + \pi^3$. Thus

$$\Delta(\pi^1 + \pi^3) \text{ is negative}$$

This was equivalent to saying that the following forces acted in the structure,

a 1-force of repulsion, at N-level
a 3-force of repulsion, at N-level,

on the simplices listed above. But at (N-1)-level this was equivalent to much higher-order forces of repulsion - over the wide range of food, drink "vertices" existing at that level.

In a similar way it is possible to examine all the items in the Accounts and to interpret their roles in the filtration/grading structure. For example, in the Academic Departments at N-level, over the same time period we obtain

a 1-force of repulsion
and a 3-force of repulsion

on simplices whose vertices are N-level groups of staffing grades. A study of Schools Administration expenses gives changes $\Delta\pi$ at the (N+1)-level, etc..

B3.1 Rules for Success in the Committee Structure

From our discussion in B2.4 it would seem that for an item of business to be accepted and passed through into action it should avoid being, for example, an (N+1)-level issue when the N-level structure S(N) contains any q-objects. This is because such business is almost certain to be decided by the pseudo-committee(s) who occupy the q-holes. In a similar way, the issue, if it is an N-level item, should as far as possible be a p-simplex (in $K_B(C)$ with $p < q$, when q denotes the least object-dimension in $K_B(C)$. This is because it will thereby be "underneath" the level of any pseudo-committee operating in S(N). As noted in section A this latter situation is structurally impossible if there exist 0-objects ($q = 0$) for there do not exist any items of business which are (-1)-simplices in $K_B(C)$.

At the same time the success of a proposal in the structures $K_B(C)$, $K_P(C)$, etc., will depend on its being injected at a sensible place - near to the decision-making point (committee). This is to ensure that the proposal is not drowned by the noise of the traffic at that level.

The overall conclusions from all these points can be expressed as follows:

To succeed in politics:

(I) inject (N-k)-level business into N-level Committees

To thwart in politics:

(II) refer (N+k)-level business to N-level Committees

Of course, under II, if there are no relevant N-level Committees it is essential to invent them - even if they must be called Working Parties. This then provides a path (in the abstract geometry) for the (N+k)-issue to play the part of traffic. <u>Only a q-object can stop traffic</u> - by inducing a decision.

<u>Case Study I</u>

A proposal in 1970/1 was "to install telephones into each flat (floor) of the residential towers" on the campus. The arguments "for" were

(i) increased security for tenants, following some intrusion,

(ii) increase service for tenants re contact with "outside" (N+3),

the arguments "against" were

(iii) installation and running costs would increase rents beyond a tolerable level,

(iv) possible "abuse" of contacts with (N+3) - outside world.

This item of business was clearly at the (N-1)-level (see page 153). It was injected into SOCIAL/POLICY Committee (which in those days was called the Student Affairs Committee) and through that to SENATE from whence it was referred to FINANCE Committee. The route can be described as follows:

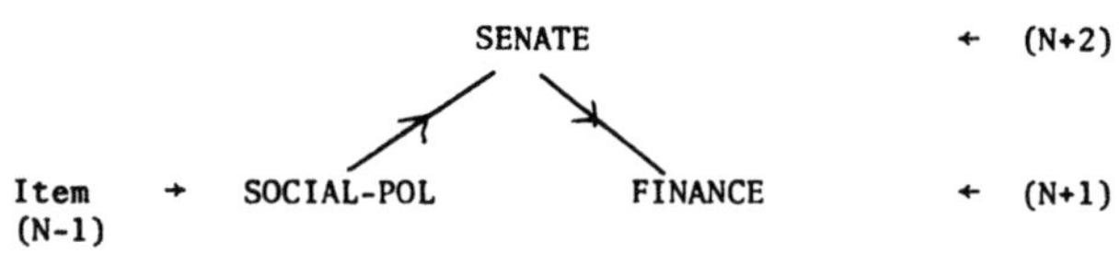

The objections re cost were met at FI and, after study, found to be unsupportable. The decision was finally implemented in 1971/2.

At SOCIAL/POL the discussion was about the above (N-1)-matters ((i) and (ii)); at FINANCE the discussion was about the pattern π_N (and its partitioning, see B3.0); at SENATE the discussion was about the (N+1)-item of Social Amenity provision.

By Rule I, the operation was quick and efficient. It was primarily made possible by SOCIAL-POL being willing to accept the (N-1)-item, avoiding the "thwarting line" of referring it down to an (N-2) Working Party.

Case Study II

In 1970/1 the proposal "to initiate changes in the appointment and allocating of Advisers to undergraduates" was accepted by SENATE and acted upon in 1971/2. These proposals amounted to a new organisation for finding, funding and allocating Advisers (members of the academic staff appointed to counsel and advise undergraduates on their personal and academic problems). They were fairly radical since they took into account the realities of a situation in which all academic members were not equally good as would-be advisers. This meant that if the quality of the service was to be improved then the quantity must fall; hence fewer but better advisers.

The proposals had overtones at various levels, such as:

N+3 Advising service gives feedback to "outside", schools, etc..
N+2 Advising service is non-academic concern of SENATE and COUNCIL.
N+1 The Schools' responsibility for undergraduate progress.
N A Departmental concern for 2nd/3rd year students.
N-1 Academic Staff sense of concern for students' problems.
N-2 Individual Adviser and the individual student share community vertices.

It was presented as an issue defined at the levels (N+1), N, (N-1) to Senate at the (N+2)-level.

By Rule I it was highly likely to succeed - and was in fact implemented in October 1971, remaining in action to this day (March 1974).

(Ref: SENATE Minutes 2257-62 and General Cttee of Senate, Minute 1047-64).

Case Study III

The idea that a Director of Physical Recreation (or a Sports Officer) should be appointed on the Campus had originated from the Students Union over a number of years, after the first days of the new University. We regard the issue as an N-level item of Business, covering as it does the (N-1)-items of athletics, ball games, etc..

It was formally introduced in May 1970 to the Student Affairs Committee (which became the SOCIAL-POLICY Committee in 1971) - at the (N+1)-level; according to Rule I. It was then forwarded to FINANCE by SENATE (M 2030) which expressed concern about the financial implications - thus relegating it to an (N-1)-partition problem for π_{N-1} (see page 158). This effectively meant that Rule I ceased to apply in FINANCE; the idea was blocked thereby and the proposal joined the traffic round the q-objects in $K_p(C)$. This meant that it was quite naturally put on the agenda for the Working Party on Social Planning which gave the traffic an extra push, recommending that the university should launch an appeal for funds to finance amenity services generally (notice that this (N+1)-Working Party was energising (N+1)-traffic around the objects; "amenity services" is the higher level business to "specific amenity service"). This endangered the proposal (N-level) by embedding it in (N+1)-business and making it traffic in the structure. Thus Rule II was in operation.

But the proposal succeeded, illustrating Rule I, by its supporters on SOCIAL-POL (January 1972) forwarding it directly to the APPEALS Committee (a committee of COUNCIL) at (N+1)-level (and even at (N+2)-level if we regard the Council as at (N+3)-level, at least in its search for funds outside).

Thus the conditions for Rule I applied and it came back to SENATE with tacit approval of the pseudo-committee(s) who could not now find any traffic lane for it. Recommendation accepted in December 1971 (General Committee of Senate 1227).

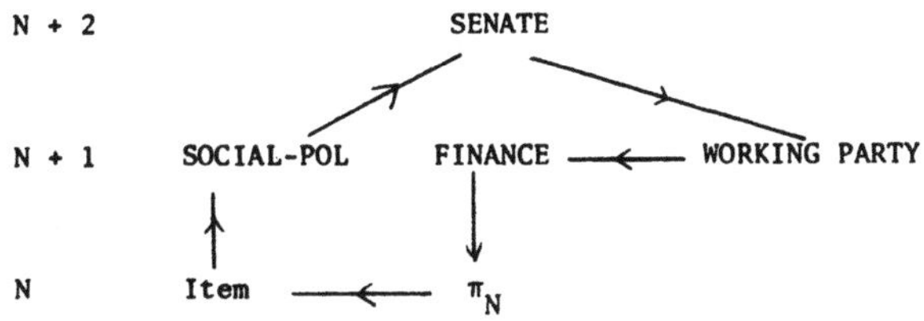

Rule I tried; Rule II applied

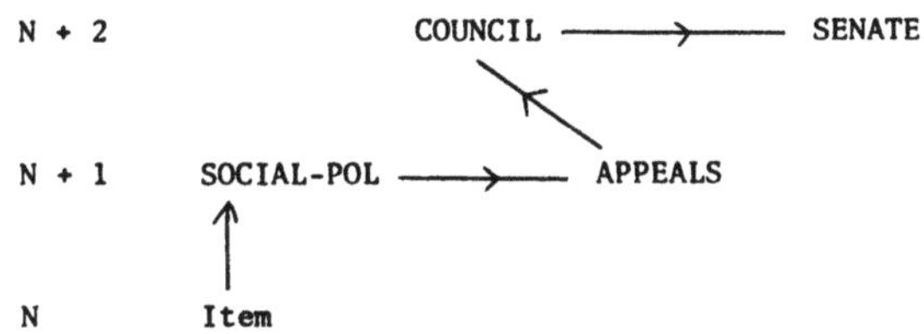

Rule I tried; Rule II avoided

B3.2 Social Policy - A Case Study in S(N+1)

Whereas the University soon found that academic matters, falling in the range of (N-1) to (N+2), had perforce to be settled quickly and consistently - and to achieve this was the prime function of School Boards (at N+1) and Senate (at N+2) - it also soon found that non-academic matters, from (N-2) to (N+2), posed a much more difficult problem. This is illustrated by most of the discussions centering on Student unrest (and which we shall examine in some detail in later sections and also by the formal action which has been taken, from time to time, in response to criticisms and suggestions from various sections of the Staff. It is in this latter context that we now look at the action on Social Policy Planning which has been going on since 1970.

In February 1970 SENATE agreed to set up a "Working Party on Social Planning", at the suggestion of the Vice-Chancellor, to study and make recommendations on "long term planning about social policy and the related physical environment". This Working Party was set up in March 1970 and had the following membership: Vice-Chancellor, Pro-Vice-Chancellor, Dean of Students, 3 members of the academic staff, 3 student members. This Working Party produced a Report in June 1971 which was accepted by SENATE and COUNCIL and which led immediately to the setting up of the proposed SOCIAL-POLICY Committee and the Wivenhoe House Committee.

In March 1973 there were criticisms of the decision-making structure in the area of Social Policy by some members of the academic staff and this led to SENATE (March 1972) setting up a new "Working Party on Arrangements for Social Planning" with the Pro-Vice-Chancellor as Chairman plus a membership of Vice-Chancellor, 7 members of the academic staff, 4 student members; the Chairman of Student's Council was subsequently co-opted. This produced a "Consultative Document on Arrangements for Social Planning' in May 1973 and a Report to SENATE in December 1973. This Report was referred to Steering Committee of Senate and was also discussed by SOCIAL-POLICY Committee in January 1974. In March 1974 a final recommendation to SENATE was ready, (present position).

The main proposals which emerged from the Working Party on Social Planning in June 1971, together with an indication of their hierarchical positions were as follows

Housing Service	Level
The University should continue to purchase and convert convenient properties when the costs involved allow rents to be charged which students can afford.	N - 1
The University should provide new purpose-built student accommodation in the form of self-contained flats, grouped in low-rise buildings which accommodate some 200-250 students.	N
The University should finance new student accommodation, either by borrowing money on the open market or by negotiating a 'package deal' with a contractor.	N + 3
Additional on-site building should be in strict accordance with an overall development plan.	N + 1
Particular attention should be paid to the housing needs of married students.	N - 1
The University should encourage any proposals by the students to build and manage their own accommodation.	N - 1
Health Service	
The Health should be moved as soon as possible into more accessible, preferably purpose-built accommodation.	N
Catering Services	
The University should fully investigate the proposal for a large catering unit which could provide a variety of catering facilities before any additional catering accommodation is provided.	N + 1
Retail Services	
The University should offer as wide a variety of retail services as possible, and should encourage potential retailers, including Students' Council.	N
The Arts	
The University should, as circumstances seem appropriate, make submissions to trusts and foundations, and, if necessary, launch a public appeal to provide a purpose-built arts centre.	N + 3

Level

The University should establish a Committee with the responsibility of promoting the arts. N

The University should encourage links with corresponding bodies in the region so as to share facilities and promote the arts in the region generally. N + 3

Physical Recreation

Regard being paid to competing claims on the University's resources during the next quinquennium, the Social Policy Committee should be asked to consider the Working Party's suggestion that a Director of Physical Recreation be appointed as early as possible in the quinquennium. N - 1

The University should construct, in the first instance, a sports hall, and, if and when funds became available, a swimming pool. N - 1

The University should keep under review the use of sports pitches, including their use by outside organisations, and any additional pitches should be provided as far as possible adjacent to the University. N - 1

Consideration should be given during the quinquennium to extending the sports pavilion, and to the possibility of providing covered ways between the changing accommodation and the social area. N - 1

The lighting of the Dri-Pla area should be extended. N - 1

Amenity Accommodation

As part of a general regrouping plan there should be from 1st October '71 a first stage of regrouping and centralising amenity accommodation (accommodation for billiards, bar billiards, table tennis, darts, a quiet room/reading room). N

Consideration should be given by the Social Policy Committee to the management of some amenity accommodation by students. N - 1

Wivenhoe House

The social amenities at Wivenhoe House should continue to be open to all members and employees of the University, no change having been proposed in present University policy with regard to usage of social amenities.

A Wivenhoe House Committee consisting of a member of staff as Chairman

Level

and four further members of staff and four students, should be established, initially as a committee of the Senate, whose terms of reference would include

(a) Determining the conditions under which persons who are not members or employees of the University might use Wivenhoe House; N + 1

(b) to consider whether Wivenhoe House might be established as a club and if so, after sounding opinion throughout the University, to bring forward proposals;

(c) advising the Senate and the Council, generally on the use of Wivenhoe House, and on the organisation of its amenities, with a view to increasing the use of the House.

The bar in Wivenhoe House should be moved (with consequential rearrangements), during the summer vacation (if possible) to the present Dining Room and part at least of the existing furniture replaced at an estimated cost of £3,500. N - 2

The long-term development of Wivenhoe House requires an architect to advise on some implications; he should be appointed in time for him to be consulted on the alterations to the bar. N + 2

The Wivenhoe House outbuildings are counted by the University Grants Committee in the University's stock of serviced accommodation; having regard to their suitability, for future scientific usage the present accommodation and services should, if altered, be capable of quick and economical reinstatement. Within this policy the outbuildings should, until they are required for scientific needs, be used to provide facilities for the arts. N - 1

The management of those sections of Wivenhoe House allocated to the arts should be the responsibility of the Arts Sub-Committee of the Social Policy Committee. N

Play Group Facilities

The University should as far as possible make available suitable accommodation to meet the growth in demand for play group facilities. N - 1

The University should keep under review the question of safety of children and should draw to the attention of parents of children in the play group the problems of safety in the University for young children even when accompanied. N - 2

Transport and Car Parking	Level
The University should, after a trial period, review the traffic flow following the opening of the West Boundary Road.	N - 1
The University should, after a trial period, review the new bus services to the University.	N - 1
In order to preserve the amenities of the Park, the University should consider restricting the use of cars in the University by certain categories of student.	N - 2

Future Committee Structure for Determining Social Policy

The terms of reference of the Student Affairs Committee should be extended so that it is responsible to the Senate for the social policy of the University. It should be renamed the Social Policy Committee. Its membership should consist of the Dean of Students as Chairman together with eight members of academic staff and eight students. Its terms of reference should be N + 1

> 'to advise and report to the General Committee of the Senate on the social policy of the University, in particular policy relating to the welfare, residence and health of students, and the provision of social and recreational facilities'.

The Social Policy Committee should initially have only two sub-committees as follows, their membership to include members of the non-academic staff:

(a) an Arts Sub-Committee, whose terms of reference should be: N

(i) to advise on and make recommendations for the encouragement and development of art, drama and music within the University;

(ii) the acquisition and performance of works of art;

(iii) the provision of facilities for rehearsal and performance and of workshops and studios.

(b) a Physical Recreation Sub-Committee whose terms of reference should be as for the present Physical Recreation Sub-Committee of the Student Affairs Committee, namely N

> 'to consider and make recommendations on matters relating to the provision and use of University facilities for physical recreation.'

The March 1974 recommendations, from Steering Committee following the Consultative Document from the Working Party on Arrangements for Social Planning, amounted to the following:

The establishment of two Committees,

(1) a Social Policy Board, at (N+1)-level,

(2) a Planning Committee of Senate, at (N+1)-level

together with a suggested scheme for managerial responsibility in matters which are essentially "short-term". The membership of (i) and (ii) would be as follows

Social Policy Board:

Vice-Chancellor, Dean of Students (Chairman), Proctor ex-officio

1 academic member of COUNCIL who is also a

member of FINANCE (appointed by SENATE)

8 academic members of staff (variously specified)

8 students and 5 others.

Planning Committee;

Vice-Chancellor, Pro-Vice-Chancellor, Deans (5)

3 students and 3 others (variously specified).

In addition the Social Policy Board would possess an Arts Sub-Committee, and the proposals suggest that

(a) "major issues and long-term planning" be the business of the Planning Committee,

(b) "mid-term planning (typically for the following academic year)" be the business of the Social Policy Board,

(c) "short-term planning and management" be the business of managerial staff.

(d) ACAD/PLAN Committee and SOCIAL-POL Committee of SENATE to be dissolved and the terms of reference of the two new committees listed as follows:

<u>Planning Committee of Senate</u>

1. To make recommendations to the Senate on all major or long-term matters of academic or social planning, and their co-ordination.

2. To review, from time to time, the terms of reference of the Boards of Schools and the Social Policy Board, and make appropriate recommendations to the Senate.

Social Policy Board

1. Subject to constraints laid down by the Senate, to make recommendations to the Senate on any matter of policy under the headings below, save that the Steering Committee of Senate may resolve that a particular issue is a matter for the Planning Committee.

   (a) Regulations relating to student residence,

   (b) General regulations, primarily concerned with student welfare, as defined from time to time by the Senate,

   (c) Rules and procedures relating to Student discipline,

   (d) Annual policy on allocation of residential places to students, and rents for such places,

   (e) Any matter primarily concerned with the Adviser service,

   (f) Annual policy on provision and use of amenities,

   (g) Social implications of any decisions taken with the University,

2. On the recommendation of the Chairman, to approve the annual requests for allocation of funds to the Chairman for disbursement in connection with the items below:

   (a) Cost of minor building alterations for social purposes,

   (b) Facilities for amenity accommodation and physical recreation,

   (c) Costs of embellishment of buildings and their immediate surroundings,

   (d) Costs in connection with display of works of art, theatrical productions, and musical performances,

   (e) Grants to the Catering Committee for uneconomic services justifiable as a matter of social policy,

   (f) Any other itme agreed by the Senate,

3. To approve the recommendations of the Arts Committee of the Social Policy Board.

4. To resolve that the Chairman bring to the attention of the Vice-Chancellor any aspect of University business which is conducted at a managerial level and which is of concern to the Social Policy Board.

5. To advise the Chairman on the allocation of funds at his disposal, subject to the constraints laid down by the Senate or the Council, (whether in the form of earmarking of funds, resolutions on policy or terms of reference of the Social Policy Board).

Finally, we give the proposed management scheme for the short-term issues.

MANAGERIAL ARRANGEMENTS

Vice-Chancellor
and
Registrar

Areas	Health	Housing	Physical Recreation	Environment	Catering
Manager	Medical Officer	Student Accommodation Officer	Director of Physical Recreation	Estates & Planning Officer	Catering Officer

NOTES: (a) Each Area would be able to establish such support groups as it wished but they would not be sub-committees.

(b) The Arts would meantime be the responsibility of the Arts Sub-Committee of the Social Policy Board.

The Working Party on Social Planning (Feb. 1970) resulted in the formation of a new SENATE Committee, viz., the SOCIAL-POLICY Committee, which replaced the previous Student Affairs Committee. This SOCIAL-POL Committee was (N+1)-committee, and has been discussed in previous sections. But it failed to meet the latent Staff (not to mention Student) objections about "participation and the importance of "social" vis-a-vis "academic" issues. This must have been because, as noted in earlier sections, the formal committee structure in the university, possesses the "citadel" property, shown in the diagram.

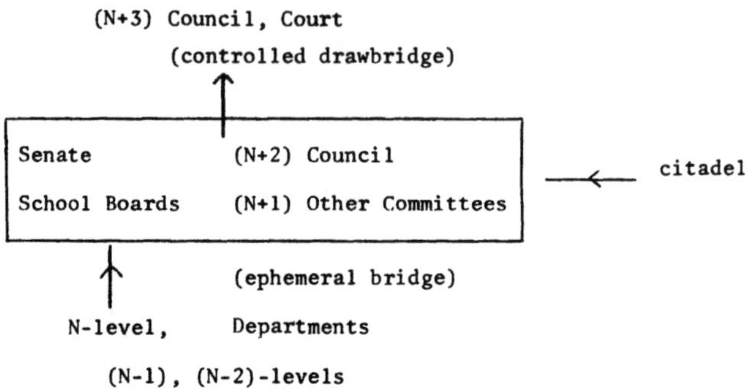

This meant that the changes introduced into the structure were "palace revolutions" inside the citadel. Since most of the Social Policy issues have significant repercussions (and indeed origins) at (N-1) and (N-2) the multi-dimensional "moat" around the citadel ensures that "the University" is experienced as an object by individuals in (N-1) and (N-2). Rearranging the structure inside the citadel is unable to create connectivities across that moat.

It is not surprising therefore that, by March 1973, there occurred yet another Staff attack on the citadel re Social Amenities/Planning. This took the form of a plea that the academic life of the community should not be divorced from its social life (opportunities in time and space for students to meet and to discuss their roles - presumably at all levels from (N-2) to (N+3). This led to the setting up of the Pro-Vice-Chancellor's "Working Party on Arrangemnts for Social Planning".

The result of this was the proposal (current March 1974) for another palace revolution within the citadel. The (N+1)-level Social Policy Board is an enlarged SOCIAL-POL Committee whilst thePPlanning Committee (replacing ACAD-PLAN) not only is to deal with the old academic planning but also the "long-term" social planning issues. Each of these new Committees is to be at the (N+1)-level - although most of the social planning of the Planning

Committee is to be concerned with (N+2) and (N+3) issues. Thus the stage is ideally set for the application of Rule II (page 161).

In addition to these committees the (N-1), (N-2) issues (usually associated with short-term) will be the concern of management (not committees) - which inevitably means Administrative (as opposed to academic) control. Thus the stage is ideally set for reference of all social policy issues under Rule II - if the pseudocommittees feel it desirable.

The situation is that of traffic, as in the scheme shown below.

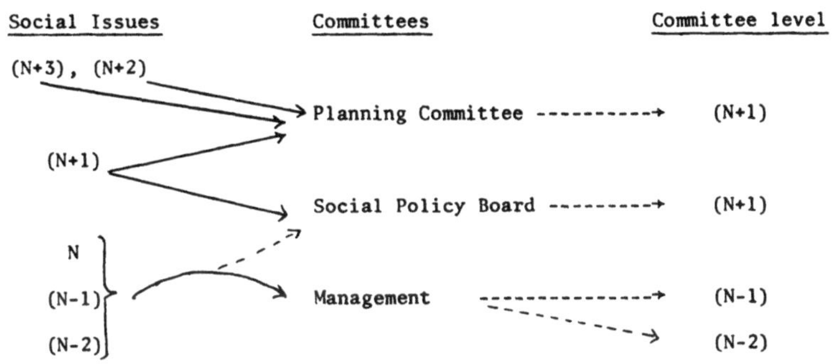

Since Management is really Administration (at (N-1) and (N-2)) it is outside the citadel, and consequently unreachable, in the formal committees. Hence the new proposed scheme must mean,

(i) The citadel remains isolated
(ii) The SOCIAL-POL Committee is further emasculated
(iii) Conditions for applying Rule II are greatly improved
(iv) Applying Rule I is made more difficult by the Managerial Structure
(v) Management is administrative and untouchable from the citadel
(vi) The influence of the pseudocommittees (inside the citadel) and of administrative q-objects (v. section B2.5) is greatly increased.

All of these deductions suggest that it is likely to be increasingly difficult for individual (academic) staff and students to play an influential role in the community.

B3.3 Wivenhoe House - a Case Study in Committee Traffic

Wivenhoe House is the 18th Century house, in the centre of Wivenhoe Park, which acted as a focal point for the new University of Essex when that began in 1964. It is remote from the rest of the university buildings and architecturally quite distinct (see Section B1.0). In 1963/4 a number of outbuildings were constructed around the original House to be used as temporary accommodation for academic and administrative purposes. At the same time some of the old House was modified to provide a restaurant, bar and common rooms. Naturally, it offers an attractive setting for the entertainment of visitors, (N+3), to the university. But on the other hand it has always been a financial burden to the university, not only because of a deficit on the catering service but also because of the very high cost of maintenance, security, and repairs for the building.

These facts are part of the background to the lengthy discussion in the university on the question of "what are we to do with Wivenhoe House?" The following account is an attempt to place not only the discussion but also the action (if that is not too strong a word) which has taken place over this question in the framework of our mathematical language of structure. Because The Question is so very difficult to answer, within the financial constraints imposed on the university, it is not surprising that the story is compatible with a high degree of indecision on the part of various committees (and of the implied <u>pseudo-committees</u>) which have considered it. Indeed, in a structure which is dominated by q-objects (and associated pseudo-committees) it is clearly the responsibility of these pseudo-committees to arrive at a decision. This is to be done, of course, by giving guidance to the many actual committees/Working Parties set up to consider The Question. If this is not done then we can see that the Question merely becomes traffic (and probably noise) on the abstract structure and indecision becomes equated with the generation of traffic noise in that abstract space.

The first thing to notice is that the Question : What are we to do with Wivenhoe House? is related to all the hierarchical levels of Business from (N-2) to (N+3). It is therefore equivalent to six kinds of questions, as follows:

(N+3) : What to do with WH vis-a-vis its function in bringing the university to the outside world (primarily via its architecture, history and entertainment role),

(N+2) : WH in its relation to the responsibilities of SENATE and COUNCIL (overall academic and social affairs, quinquennial plans etc.),

(N+1) : WH in its relation to (N+1)-architecture, social amenity on the campus,

N : WH and the N-level business items such as catering, sports, arts, recreations and Departmental academic uses,

(N-1) : WH and matters of detail, such as restaurant service, use of specific floors/areas for academic, administrative or recreational use,

(N-2) : WH as a set of fine detail; specific rooms and their uses, architectural features both internal and external; the fabric and surrounding gardens; details of menus available in bar and restaurant; hours when available; security details.

An important part of the difficulties found in discussing The Question must certainly lie in the failure to use an hierarchical filtration such as the above. To deal with it in the framework of the university's formal committee structure naturally requires that it should be referred to every possible level in the hierarchy.

But here we see an immediate difficulty since (compare B0.0 and B3.2) the formal structure shows the "citadel" property:

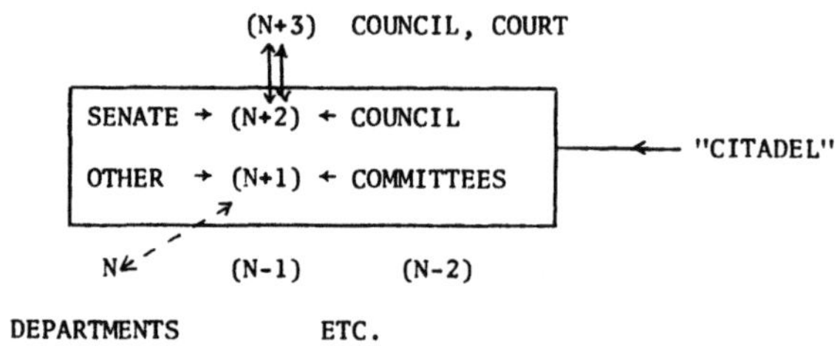

Now we have seen in earlier sections that the Citadel contains a number of Shomotopy q-objects and that this means that most of the decision-making responsibilities fall onto the pseudo-committees which perforce occupy these q-holes. If the decisions (on any of the six classes of questions) are not forthcoming then it is inevitable that the questions must begin to move in the geometry (whose points are committees) between the holes, thus becoming traffic. Paradoxically it would seem equally inevitable that the noise of this traffic can so distort the geometry that communication between the objects is seriously affected. This is likely then to make decisions (by the objects/pseudo-committees) even more difficult to arrive at. This then results in the q-objects giving more stimulus to the traffic to keep it moving - and so noisier, etc..

In the first instance, to facilitate the traffic flow, it is probably helpful to fill up the inter-object spaces by creating more points (Committees). Thus, this vicious circle of indecision, inevitably begins by "setting up a committee to consider The Question". In this specific instance we are able to see the cycle of events in the following programme.

The committee set up in March 1970 was known as the Committee on Future Use of Wivenhoe House. It reported in June 1971, to SENATE, when its recommendations were largely those already listed on page 165, section B3.2. We notice there that the questions which were implied fall into the hierarchical range of (N+2) to (N-2).

In October 1971 the SENATE set up a new Wivenhoe House Committee to answer The Question. The record of its minutes during the period October 1971 to June 1973 shows the following breakdown into hierarchical levels.

Sequence of items	Level
Consideration of WH as a Club	N
Approx. £2500/annum needed via subscriptions	π_N (Finance)
Subscriptions discussed as affecting individual members	(N-2)
Rearrangement of ground floor accommodation agreed	(N-2)
Subscriptions discussed, proposed referendum of university	(N-2)
Alterations to bar, decor, furniture, with Architect present	(N-2)
Response to referendum (8% students, 10% overall)	
Deduced "no clear mandate given"	(N-2)
Approach to Chairman from outside Brewery re: converting Wivenhoe House into an hotel, encouraged	(N+3)
Alterations to bar to be deferred	(N-2)
Sub-committee formed to negotiate with Brewery	(N+3)
Study of room sizes showed unsuitable for use as a Students Union	(N-2)
Possibility of attracting "some educational institution" to use WH	(N+3)
Reported that outside Brewery no longer interested	(N+3)
Proceeded to study details of costs involved in developing the use of WH	π (Finance)

(Sequence of items) cont. **Level**

Bar alterations to be commenced (N-2)

Estimated repairs approx. £70,000 over next 5-7 years $^{\pi}$N-2

Estimated running costs £9,000/annum (excluding catering loss of £4,000/annum) $^{\pi}$N-1

Reported that outside bodies interested in development of WH (N+2)

Leaflet produced to publicise "facilities available at WH" and signs to be erected showing the way to restaurant/bar (N-2)

Reported that SENATE had reaffirmed that WH be developed as a (N+2) social centre and did "not wish to rule out expenditure of the order of £81,000 (capital) + £13,000/annum" (N+2)

Bar alterations were now completed (N+2)

Committee discussed whether it should continue as a committee of SENATE and what sub-committees it might set up with a view to promotion of the use of WH.

Decided to continue in existence as a SENATE committee "to advise the SENATE <u>and COUNCIL</u> generally on the use of WH".

Set up a sub-committee, the Wivenhoe House Social Committee, to (N) be responsible for the active promotion of WH (N-2)

Agreed to defer any review of the long-term development plan until the quinquennial details (January 1973) were known and until (N+3) appointment of consultant and architect in Spring Term 1973.

Catering report for August 1972-January 1973 and proposed Budget for 1973-4 received. (N-1) Deficit of order of £8,000 agreed that the remedy was to reduce service (particularly waitress service in restaurant) (N-2)

In March 1973 decided to defer discussion of long-term plans until WH Social Committee had reported; also to await suggestions from (N) future consultant architect (N+3)

(Sequence of items) cont. **Level**

In July 1973 COUNCIL created a new Working Party to report via
SENATE to DEVELOPMENT Committee, not later than Spring 1974, on (N+2)
The Question, this to include 3 lay members (one as Chairman)

Reported to COUNCIL (July 1973) that the Department of the Environment
had now included WH + outbuildings in its list of buildings of (N+3)
"Special architectural and historical interest".

Meanwhile (May 1973) the WH Social Committee met to consider its
responsibilities. Chairman undertook to find out which specific rooms (N-2)
were their responsibility; noted that 'large areas of basement and
Centre first floor were already allocated and not available for their (N-2)
consideration'. Expressed concern about the fabric and repairs. (N-2)
Discussed room bookings and confusion among potential users. (N-2)

Recommended specific reductions in service (N-2)
On general lines of social development concluded "prospects of improving
the social use of WH are poor but not hopeless" N

....

The most striking feature of this list of items discussed in the various committees is the fact that 80% of them fall into either (N+3) or (N-2). That is to say, most of the discussion and decisions fell well outside the "citadel" defined by the formal committee structure.

Of these matters of business the (N+3)-items could only be settled by the (N+3)-committee, viz., COUNCIL. The (N-2)-items could only be settled by (at least) an (N+1)-committee (because there are no N-committees in the citadel), that is to say, the WH Committee itself or SENATE.

Using the 1972/3 Data file of C v DP we notice that the (N-2)-business would have involved SENATE - which is a component in the 0-object (discussed on page 150).

G2 :{AC/PL, COUNCIL, MA-ST, DISCIP, SENATE, AC/PL}

But, via C v P, the WH Committee was shomotopically equivalent to SENATE (and also to SOCIAL-POL). Thus, business (viz., The Question) through the WH Committee was naturally traffic round the 0-object G2.

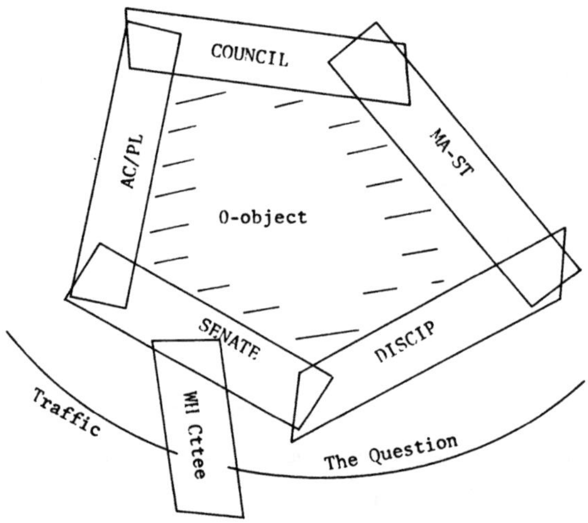

Figure 11. The Question as noise

A decision on The Question (at either N-2 or N+3) in any of the issues listed above could have been reached only by a pseudo-committee (on G2) directing the traffic into the hole (bringing it to rest in (say) SENATE or COUNCIL). But this was avoided (in effect) at certain stages by the deflecting device of creating new sub-committees (such as the WH Social Committee or the Council's Working Party (July 1973)). The geometry was therefore altered via the following scheme, and The Question remains (March 1974) traffic, undecided.

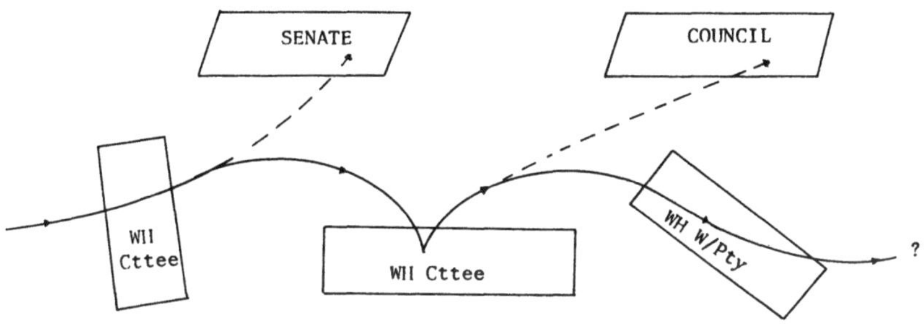

Figure 12. Keeping the traffic moving

B4.0 Student unrest - Stress in the Structure

In recent years this university, and most other universities at home and abroad, have experienced varying amounts of student "troubles" - in the shape of mass rallies, demonstrations, protests, sit-ins, strikes, demands for change/reform in most areas of university life. An important aspect of the recurring demands has referred to participation by undergraduates in the "decision-making-processes" of the university community - and this has clearly been linked with the concept of a university qua community, and the struggle to achieve some idealised version of that concept.

During the course of this research study some of the undergraduates at the University of Essex, backed at various stages to various degrees by the Students' Union, undertook a sit-in (or occupation) of certain areas of the university buildings as a means of drawing the attention of "the authorities" to specified complaints and demands. Their actions, and those of the authorities, in the ensuing struggle are summarised below. What chiefly interests us here is how these actions reflect the existence of stresses and strains inherent in the Structure, at various hierarchical levels. The relevance of the Structural Language is therefore suggested by the form and content of the table-of-events below.

Date	Event	Structural remarks
15.11.73	A Student 24-hour strike called by the National Union of Students (NUS). Picket lines outside lecture theatres. Some Staff complained that their lectures were disrupted by intrusion of militant pickets. The strike was in support of a national campaign for improved student grants.	Student personal problems (at N-2) placed on a national stage (at N+3). Strike action aimed at (N-1) and N-level features of university life.
19.11.73	A number of administrative offices were occupied by students, who proclaimed that their struggle was against "the administration". Some confidential papers re. students and staff were copied and made public. The Demands were the following: 1) An immediate 15% reduction in catering prices, with no staff redundancies. 2) Firm guarantees that catering and accommodation facilities will be expanded to cope with the projected increase in student numbers. 3) An immediate 55% reduction in rents at Cambridge Road and The Avenue, Colchester.	The opposition to "administration" expressed anti-traffic feelings in the structure. This traffic is the life-blood of the citadel. The demands were at (N-2), (N-1) and N-levels, "outside" the citadel.

Date	Event	Structural remarks
20.11.73	The Pro-Vice-Chancellor notified the students that legal action would be taken against them if they did not vacate the premises by 12 noon. They failed to do so. A General Meeting of Students resolved the following: 1) In furtherance of its demands actively to support the present sit-in until such time as the demands are met in full. 2) To support any further measures of direct action which may at present or from time to time be necessary in order to achieve these demands. 3) To approach Colchester Trades Council and request its support for our demands. 4) To enlist the support of all Trades Union on campus. 5) To approach Tenants Associations and squatters organisations in the area, to request their support and help to discuss future joint action on the housing question. 6) To contact the Students' Union at North-East Essex Technical College to request its support and to discuss future joint action on housing, rents and grants. 7) This meeting wishes to affirm the wholehearted support and solidarity of Essex University Students' Union for the just struggle of the Greek people for the restoration of democracy, trade union rights and academic freedom. In particular the Union expresses its Solidarity with the Students of the Polytechnic Institute of Athens and their struggle against armed police and military repression. This meeting mandates the executive to send telegrams to this effect to the four main universities and main newspapers of Athens.	University authorities appeal to (N+3)-High Court for help. Notice that the student resolutions (1) to (7) contain a widening appeal to (N+3)-levels (even at international level).
21.11.73	The High Court granted the University an order for possession; this was served on the same day. Students issued a fly-sheet which (1) reiterated the 3 Demands, (2) defended their action against the University, (following Staff claims that the proper target was the University Grants Committee and/or the Government) by saying that "student participation in the formalities of UGC negotiations get bunged up with red-tape". (3) proposed that all (academic/work deadlines be waived for the duration of the sit-in.	Help comes from (N+3) to the citadel. The students begin to worry about the possibility of future disciplinary action - (N-2)-effects imposed by people who are safe inside the (N+1)↔(N+2)-citadel. Disciplinary rules assert "lane discipline" for the student <u>traffic</u> through the structure?

Date	Event	Structural remarks
21.11.73 (cont)	(4) that no disciplinary action be taken against the students for failure to attend classes for the duration of the sit-in.	
22.11.73	Students sent an Open letter to the Vice-Chancellor with the Demands 1) That a written guarantee be given that there will be NO victimisation of those in occupation either through the courts or the University Disciplinary System. <u>This demand is the pre-condition of any discussion of the others</u>. 2) That there be an immediate, across the board, reduction of 15% in catering prices, without staff redundancies. 3) Provision of catering and accommodation facilities to cope with any increase in student numbers. 4) An immediate 55% reduction in rents and accommodation charges at University owned property in Cambridge Road and The Avenue. 5) That either you or acceptable representatives appear at a General Meeting of Students and Staff called for Friday 1 pm in the LTB to answer these demands.	The new 4th Demand of no-victimisation is firmly to the fore; see previous remarks re. discipline.
23.11.73	VC reported on the situation to a special meeting of the General Committee of Senate; declared willingness to discuss the issues at any time with student representatives, but quite unable to abandon the normal disciplinary procedures. An Extraordinary General Meeting of Students (approx. 400 attending) carried a motion "that the Union refuse to make payments to the University for damage which may have occurred during the sit-in".	VC reports at (N+2)-level. Willingness to discuss at a personal level, at (N-2). Illustrates the lack of formal structure to bridge (N-2)→(N+2) gap. Students preparing to oppose finance issue, re. π_{N-2}, π_{N-1}.
25.11.73	Bulleting issued by International Marxist Group (IMG) in support of militant action at Essex as part of campaign against the "UGC and Tory Government"; a call for joint action with trades unionists and workers.	Illustrates the strong element of (N+3)-level activity in the fight.
26.11.73	VC circulated a latter to each member of Staff, repeating the history of events and rejection of "no disciplinary action" demands; "the university has a duty to ensure the essential conditions of an academic society and these include the holding of lectures and classes without disruption, the unobstructed conduct of	Statement of protection of, and definition of, the academic community, by the citadel - as - object

Date	Event	Structural remarks
26.11.73 (cont)	administrative work and the confidentiality of personal files.	
28.11.73	A General Assembly of Staff held to hear report from VC. Passed a motion calling on the students to end the sit-in. Rejected three motions of a concillatory nature, viz., (1) We wish to express our agreement with the students' original 3 Demands and call upon the University to make an adequate response to what is widely recognised as the basic justice of the students' case in regard to grants and financial problems. (2) We wish to make clear our concern about the way in which the University administration has handled the conflict, particularly in its hasty resort to legal action. There should have been greater consultations from the beginning with academic staff. It is not to the credit of administration that so little information about their actions in this matter percolated down to the staff. (3) We propose that a general amnesty be granted by the University towards any individual involved in current student actions arising from the grants campaign. We think that the institution of disciplinary proceedings against individuals is both impractical and counter productive. The issue was no firmly that of "no-victimisation".	"Calling on the students" appeals to most academic staff - illustrating the (N-2)-ties between staff and students. Some staff willing to abandon the sanctuary of the citadel.
3.12.73	A General Meeting of Students discussed and finally rejected a call to end the Occupation as from Friday 7th December.	Students beginning to disagree about tactics - because of confusion about targets/issues at various hierarchical levels?
6.12.73	Members of the students' "Strategic Committee", at 5.40 a.m., attempted an occupation of the University Telephone Exchange. The barricades and students were removed by security staff and Proctor in a resolute counter-attack by the authorities at 7.00 a.m. Students claimed that the aim of this particular Occupation was to intercept all calls to/from the Administration (only).	Attempt to sever the $(N+2)\rightarrow(N+3)$ links, but hopefully to retain $(N-2)\rightarrow(N-2)$ links with academic staff. Strong and instant self-preserving action by citadel.

Date	Event	Structural remarks
11.12.73	The VC received the following letter from the Students in Occupation: "The Occupation is only a beginning. You have not acceded to our demands. Therefore direct action will proceed next term on an increased scale until the demands are met in full. Lightening occupations, strikes and disruption, canteen boycotts, and blockades of Senate and other administrative bodies will all become a regular feature of life at Essex until we are assured of a decent standard of living and the right to determine our own destiny without fear of legal or disciplinary reprisals. We have voted to end the occupation. But we shall return next term (only a complete closure of the University can prevent that) armed with other, more powerful weapons. That mobilisation will continue. However, we also intend to join forces with that most powerful sector of society - the working class. Remember 1972, when the miners came to Essex? We intend to repeat that joint worker/ student activity with redoubled vigour in the New Year when not only the miners but other groups of workers will be launching an o offensive against this Government and its policies. Your empty threats of victimisation do not deter us. Our campaign will not cease until you meet our just demands." The Occupation was quickly ended by the students in the early hours of 12.11.73.	The tactics of sit-in had become lost in a maze of levels. The citadel-object had withstood the attack. Threats of future action were a mixture of N+3 (the miners), N+2 (Senate blockade), N+1, N (various others) etc..

....

Of the many facts to this particular sequence of events, symptomatic as they are of many similar related revolts in a variety of universities and colleges, we pick out the following as salient in the context of this Research Study.

1) Fundamental issues are raised as to the nature and purpose of a university education. It would appear characteristic of undergraduates that they develop very quickly, after admission, in their ability to be sensitive to an extensive set of hierarchical levels. After all, most of the instruction they receive, via

lectures, classes and discussion, is an extended exercise in building up a hierarchy (in our sense) of conceptual levels. This readily induces an enthusiasm for high-level political "solutions" to the lower level (individual and social) problems. It is perhaps therefore not surprising that, in a conflict such as the above, there will be those who see it entirely in terms of an idelogical (N+k)-level. It is consequently a class struggle, a national struggle, even an international struggle. The use of (N+k)-level concepts, in a student debate, makes such an appeal practically irresistible, motions are passed, enthusiasm generated for the "struggle". Later, as the action impinges on (N-2), (N-1), N-level problems it becomes more difficult for the individual student to understand the subtle relevance of the (N+k)-concepts. Eventually a sense of defeat ensures as a more or less bewildered experience - the "establishment", or the "administration" has won again.

But this sensitivity to the hierarchical filtration inherent in both academic and social/political life is possibly the essence of education, when that is not purely vocational? If the academic educators applaud its cultivation in, say, mathematics, then how can they deny it in other contexts?

2) Since the University, qua institution, is based on a formal structure which is both hierarchically filtered and multi-dimensionally connected it seems inevitable that academically promising students who become sensitive to these features will respond to the need to play a non-trivial role in the mechanism. But we have seen that when there is a large number of q-objects in that structure then that non-trivial role can become practically impossible to achieve. In this instance we can liken the situation to that of the medieval peasants trying to storm the citadel. But the action, which takes place on the campus and in its buildings, etc., is really a representation

of action in the abstract multi-dimensional space, the backcloth S(N), S(N+1), etc.. When the citadel holds fast then the action perforce must be all around it, as shown in Figure 13.

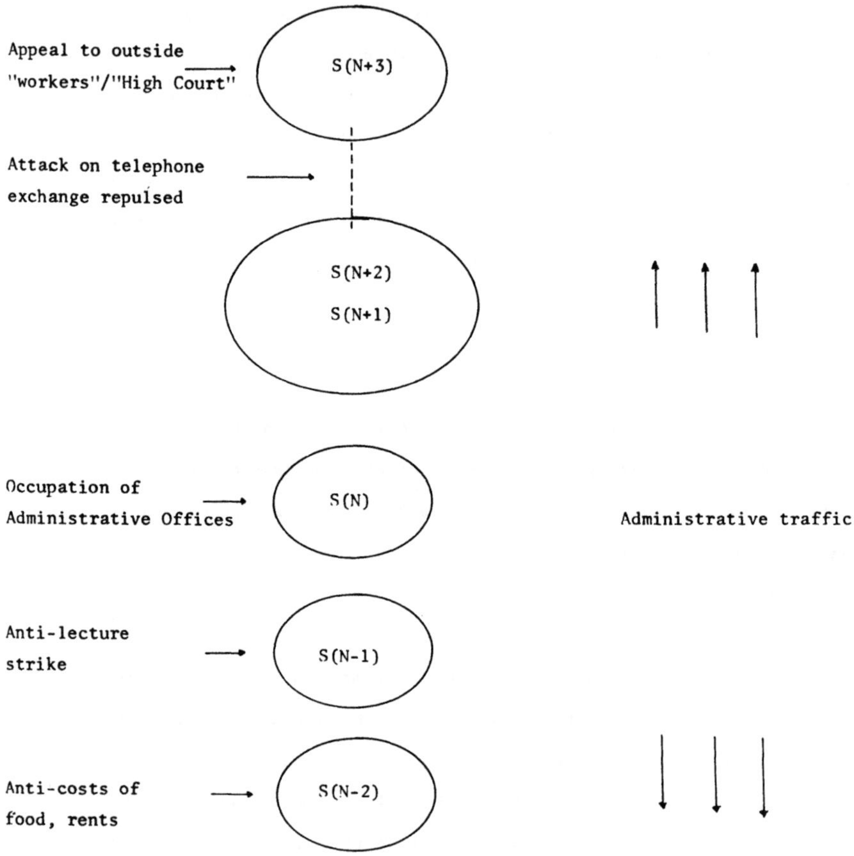

Figure 13. Student militancy on the backcloth

3) The problems facing the staff, both academic and administrative, must be at least threefold.

(a) to preserve the structure of the citadel insofar as they themselves are objects and/or traffic in that structure,

(b) to control the manifestation of basic academic aims in the structure of the university; for example, to educate the students in such a way that they can appreciate the structure and identify with it sufficiently strongly as to join in its preservation (this gives the university a meta-structural role), and

(c) to engage in research, extending the frontiers of knowledge; this places the successful academic scholars in a wider backcl th, S(N+3) which transcends that of the local university; thi in turn might weaken individual needs to preserve the local structure.

Because of (c) we would expect some numbers of the academic staff to find it difficult to constantly oppose student action/demands - militant or otherwise. On the other hand, those members of the academic staff who in effect are content to throw all their energies into purely administrative tasks are more likely to oppose such action. But even for these administrative/academics the local structure, and in particular the presence of shomotopy q-objects, must pose severe problems of identity and role.

4) In a wider context, in society at-large, it must be the case that the university is a q-object, and in that context it must be represented by its highest office holders - usually the Vice-Chancellor. Thus the VC, as the university-object, must be part of the traffic around other objects of a national kind. It therefore becomes acutely necessary for the link between (N+3) and the citadel to be maintained - for otherwise the university-object cannot function as traffic. In maintaining this link, furthermore, the university must be completely subordinate - it is a messenger, an agent of

multi-dimensional (higher) national objects. This is why, where the events/actions of a typical sit-in are reported to COUNCIL or COURT, the tone (of the university object) is entirely subordinated to outside standards and expectations. The "strong" action of the authorities must be emphasised, the number of student militant agitators must be "a small clique" who have temporarily seduced the rest. What is intriguing at the (N+3)-level is the fact that, vis-a-vis the higher national objects, the internal structure (at lower levels) of the university-object is probably irrelevant. What upsets the (N+3)-objects is the noise of (metaphorical) gunfire within the university-object - because it might mean that the university is about to cease to be an object (and therefore suitable as traffic) due to internal collapse. If that ever occurs in any university then the titular head will always be at risk vis-a-vis the national objects.

B4.1 Intuitive awareness of the Structure manifest through Interviews

In a series of personal interviews with over 60 members of the University, including academic staff, students, administrative staff of various grades, it has been possible to demonstrate the intuitive awareness of the structure which is felt by individuals. This relates the way people really talk about their community environment and their personal experience of it. is asked to fill up a questionnaire, but each is invited to talk freely about himself vis-a-vis the university.

The results are collected in the sequence below, without reference to individuals or to their formal status in the community. The Remarks are references to previous sections and/or to concepts peculiar to the overall analysis, the main entries being quotations or paraphrases of quotations from the interviewees.

On social amenities and services	References
We need more common rooms as social contact centres.	B1.2, B1.7.
We need a Students' Union Building with bars, restaurants, etc.	B12., B1.3.
Could Wivenhoe House be a Students' Union Building?	B1.3, B1.7, B3.3.
Centralised amenity facilities would attract students to apply.	B1.7.
There is a lack of control of own social lives.	B2.2, B2.3.
The restaurants cater primarily for admin/academic staff, not students.	B1.4.
There is a lack of locations where chance meetings take place.	B1.3, B1.7.
There is a need for more facilities clustered around.	B1.3, B1.7.
Having regretted coming to Essex, chiefly on social grounds.	B1.2, B1.3, B1.7.
There is a need for intermediate sized social groups.	B3.2.
I disagree with the idea of an "integrated university".	B1.2, B1.3, B1.7.
The ideal of "no them-and-us" is outgrown.	B2.1, B2.2, B4.0.
There are no shared social amenities on scale originally envisaged.	B1.4, B1.7, B4.0.
I don't know who my Adviser is.	B3.1.
The campus site is grey, but unauthorised painting of buildings is immediately removed by Security Staff	B1.0, B1.1.
I sympathise with the student view that the university is a jungle.	B3.2, B4.0.
The students need an area that is entirely theirs.	B1.2, B1.3, B1.7.
Catering, housing, amenities, maintenance should be run by small committees of academic staff and students.	B2.2, B2.3, B2.4.
The VC needs to wear a pullover and visit the Coffee Bar occasionally.	B0.0, B4.0.
The VC's notion of equality has resulted in there being no place for action.	B1.2, B1.3, B1.7.

On Committees and politics

I am not interested in standing for committees.	
At Departmental meetings one feels that decisions have all been taken elsewhere.	B2.2, B2.3.
The possibility of being on Disciplinary Committee was put to me by the Registrar and by the Academic Registrar.	B2.5, B4.0.

On Committees and politics (cont).	References
There is a barrier between students and the administration.	B0.0, B2.2, B3.1.
"Top people" must improve relations with students.	B2.2, B4.0.
The administration regards students an inanimate objects.	
The structure is highly centralised.	B0.0, B3.2, B4.0.
The administrators' interests are not the same as those of the academic staff.	traffic, versus objects?
The function of the School Boards is to slow down radical change.	B2.1, B2.4, B3.2.
The "usual channels" are not profitable.	B2.4, B2.5, B3.2.
The machinery is only effective when you have personal contact with the Finance Officer.	B3.0.
Students on committees were originally a vanguard, but what should they do now?	B2.2, B2.3, B4.0.
The power structure consists of a few people meeting in corridors.	B2.3.
People on committees are not told everything, the issues are already decided.	B2.3.
Finance committee does not share the cake, it is already decided.	B2.3, B3.0.
Committees consist of the VC plus "dead ducks".	B2.2, B2.3.
The Sit-in is an exercise in student politics.	B4.0.
The students who sit on committees are booed by the others.	B2.3, B4.0.
On catering no matter what views we express we cannot influence the decisions.	B2.2, B2.3.
I'm not oppressed by the structure because I know Albert, Douglas Robin, etc..	B2.3, B3.1.
The administration are in an ivory tower, particularly the Registrar and the Finance Officer.	B2.5, B3.2.
The VC pulls the strings.	B2.3.
The administration says, we've got the money and we know best.	B3.0, B3.1.
Committees are ineffective because decisions are thrown from one to another.	B3.2, B3.3.
The administration gives a full ear, but the ear is blocked.	noise?

The disconnection of the citadel	References
The university is authoritarian, excluding students from decision-making.	B2.3, B3.2.
I resigned from Disciplinary Committee when they applied for the High Court Writ.	B4.0.
For instance, the Physical Science,Board does not like students; this seeps through the whole system.	citadel?
Students on various committees are not listened to.	B2.2, B2.3.
The University has too many Chiefs and not enough Indians.	B2.3, B2.3.
Students do not trust structure to deliver the goods.	B2.4, B3.2.
Since thr troubles of 1968 the Administration has not learned.	citadel?
I know absolutely nothing of Senate.	citadel?
Committees are ineffective, but we need more representation on committees.	?
Students on committees is a token gesture, like the single negro in a cheap film.	B2.2. B2.3.
The Sit-in resulted in us having a place to meet and talk.	B1.3, B1.7, B4.0.
A lot of people get thwarted by the structure.	B2.1, B2.2, B3.1.
On committees students feel impotent, the patrons patronize.	B3.1, B3.2.
The Administration on committees are like a brick wall.	citadel?

On academic matters and the outside	
This is the best Sociology department in the country.	B2.0, B2.1.
We have enough applicants, we take 40% E-E because we are anti-elitist.	B2.1.
Degrees are finished.	B2.1.
My niece was warned off Essex by her headmistress.	
If we had a modular unit system many students would choose Chemistry and Physics.	B2.1.
Sociology etc. are Mickey Mouse subjects.	
We need a much higher standard of entry to the University.	
There should be a greater turnover of academic staff, tenure should be abolished.	B2.0.
The major attractions of the university are the research facilities.	

On academic matters and the outside (cont.) | References

Mathematics students on the whole are dull, they just go on doing mathematics.

The University is looked down on by Careers Masters.

My headmistress tried to dissuade me from coming.

Essex was my 5th choice, but I'm glad I came - not staying for academic reasons.

A final precautionary note: in the interviews it became clear that most people use the word "administration" to encompass the whole authoritative structure; professional administrators should not read to much into apparent aspersions contained in the above quotations.

B5.0 Meta language questions

We end this Section by posing questions about the methods and/or aims which are involved in actually using the Structural Language. This study has been concerned with a specific community, some would call it an institution (q-object(s)) in our society, but if it is relevant there then it is relevant for any social organisation which can be similarly well-defined. But it is then necessary to have a clear idea of how precisely the methods could be applied (e.g.) to a university; how to identify the problems; how to use the concepts of the language to build an institution compatible with a high level policy criterion; how to monitor changes (both actual and projected) so as to cope with future forces/stresses which will be induced by those changes; how to plan ahead so as to avoid unwanted and wasteful stress. In other words, can the Analysis help us to find a way to good government?

In the light of the discussion in earlier sections it is suggested that a practical programme for tackling these questions would need to include the following possibilities.

1) Set up a consistent set of hierarchical levels for all aspects of the social, academic, political life of the institution. This must clearly include a careful filtering of the issues (items of business) which are constantly requiring decisions.

2) Construct all relevant structures S(N) and examine their global and local geometries (via the structure vectors and the generators of the shomotopy groups).

3) Identify relevant patterns $\{\pi\}$ on the various S(N) and examine the significance of the filtrations and the partitions on each such π. This is particularly relevant to the overall use of resources such as finance, man-power, student applications, and student graduations. Thus we would need to ask the questions:
How are student drop-out numbers related to the hierarchical filtration and/or the pattern grading (partitions)?
How are the financial resources to be filtered and partitioned, and by whom?
What is the relation between the filtration of finances and the employment of staff?
What is the relation between the shomotopy groups of the Degree Schemes and the pattern of student numbers?

4) In each case of backcloth S(N), for all relevant values of N, examine the shomotopy structure and the induced administrative traffic. When does the traffic become noise? How far does the presence of q-objects act as a help or a hindrance to arriving at widely acceptable decisions? Is the number of such q-objects critical in any sense? How are the university office-holders (both staff and students) to function as q-objects in each S(N)?

5) Attempt to define the university as an object in S(N+3). Is the concept of the university as a "citadel" compatible with its function in S(N+3)? If so, where is the citadel to be located (in the hierarchical levels) and who is to be allowed membership? If (e.g.) the students are not to be allowed membership then how are their aspirations to be catered for (for example, at (N+1) and (N+2))?

6) What is the nature of the traditional idea of student discipline? Is it inevitably linked with the existence of an exclusive citadel? To what extent are the conflicts associated with discipline determined by the non-academic backcloth in S(N)? Since S(N), for any N, is in a multi-dimensional space it is possible for different citadels to exist and not to meet (or the traffic round them not to clash). Is a traditional Students' Union (building) an example of a citadel, providing q-objects for student officers to occupy?

....

B-References

1. ATKIN, R.H. — Urban Structure Research Project, Research Report IV, A Community Study - The University of Essex,(1974).

2. UNIVERSITY OF ESSEX — Committee Minutes:
Senate, Council (all meetings 1964-73), Finance, Development, Catering, Landscape, (1970-1973), Social Policy, Academic Planning, School Boards (of Mathematical Studies, Physical Sciences, Social Studies, Comparative Studies), Wivenhoe House (various) (1970-73)

Audited Accounts 1971/2 and 1972/3
Various files/documents on Housing, Catering, Security, Staff, Students, Estates & Planning, Maintenance and repairs.

APPENDIX

Computer Output

SHOMOTOPY OBJECTS IN THE UNIVERSITY STRUCTURE

Data file	Complex	q-value	$(S_q, *)$ generators
S v L2	K(X)	0	G1 : {L1 L7 L2 L15 L12 L3 L24 L1}
SOC v L	K(X)	0	G1 : {L1 L11 L3 L9 L7 L1}
			G2 : {L4 L6 L9 L3 L7 L24 L15 L4}
SA v L	K(X)	3	G1 : {L1 L7 L4 L9 L6 L18 L1}
SA v L	K(X)	2	G1 : {L1 L7 L4 L9 L6 L18 L1}
D v L	K(Y)	0	G1 : {AR GV CA MA CC MN AR}
D v L	K(X)	0	G1 : {L1 L3 L2 L11 L5 L9 L1}
			G2 : {L2 L11 L5 L9 L14 L2}
C v P	K(Y)	4	G1 : {AC/PL, CHAIRS, MA/ST, CPTG, SENATE, AC/PL}
C v P	K(Y)	3	G1 : {AC/PL, COUNCIL, MA/ST, BURROWS, SENATE CHAIRS, AC/PL}
			G2:: {AC/PL, COUNCIL, SOC/ST, DISCIP, SENATE, CHAIRS, AC/PL}
			G3 : {AC/PL, COUNCIL, SOC/ST, BUXTON, SENATE, CHAIRS, AC/PL}
C v P	K(Y)	2	G1 : {AC/PL, PHYS/SC, BURROWS, SOC/ST, CPTG, CHAIRS, AC/PL}
			G2 : {AC/PL, PHYS/SC, BURROWS, COMP/ST, LAT/AM, CHAIRS, AC/PL}
			G3 : {AC/PL, PHYS/SC, BURROWS, COMP/ST, LIBRY, SENATE, AC/PL}
			G4 : {AC/PL, PHYS/SC, BURROWS, COMP/ST, BUXTON, CHAIRS, AC/PL}
P v C (XR = 3)	K(X)	3	G1 : {AC/PL, STAFFING, CHAIRS, CPTG, MA/ST, SENATE, AC/PL}
			G2 : {AC/PL, STAFFING, CHAIRS, SOC/ST, DISCIP, SENATE, AC/PL}
G v D72	K(Y)	4	G1 : {MRGS, CLPS, TSNO, TECO, CLCS}
G v D73	K(X)	4	G1 : {AC FI MN CC}
DPDS72	K(X)	0	G1 : {ECON/GOV, GOV(CS), GOV/SOC(CS), SOC(CS) GOV/SOC(SS), ECON/GOV}
			G2 : {MA/SOC, SOC(CS), GOV/SOC(SS), ECON/GOV, GOV/SOC(CS), MA/SOC}

Data file	Complex	q-value	$(S_q, *)$ generators
DPDS73	K(X)	0	G1 : {ECON/GOV, GOV(CS), GOV/SOC(CS), SOC(CS), GOVE/SOC(SS), ECON/GOV}
			G2 : {HIST/GOV, GOV(CS), GOV/SOC(CS), SOC(CS), GOV/SOC(SS), HIST/GOV}
DPDS74	K(X)	0	G1 : {GOV(CS), GOV/SOC(CS), SOC(CS), GOV/SOC(SS), ECON/GOV, GOV(CS)}
			G2 : {GOV(CS), GOV/SOC(CS), SOC(CS), MA/SOC, GOV/STATS, GOV(CS)}
			G3 : {GOV(CS), HIST/GOV, GOV/SOC(SS), SOC(CS), GOV/SOC(CS), GOV(CS)}
			G4 : {SOC(CS), GOV/SOC(SS), ECON/GOV, MA/ECON, MA/SOC, SOC(CS)}
			G5 : {MA/SOC, TH/PHYS, CHEM/PHYS, MA/CHEM, MA/SOC}
C v DP SV = 3	K(Y)	0	G1 : {AC/PL, CHAIRS, COMP-ST, DISCIP, SENATE, AC/PL}
			G2 : {AC/PL, COUNCIL, MA-ST, DISCIP, SENATE, AC/PL}
C v DP SV = 1	K(Y)	5	G1 : {BURROWS, CHAIRS, DISCIP, ED-TECH, BURROWS}

Visual sets

N-level

Set v		{Views}
V_1	-	Square 4
V_2	-	Square 3
V_3	-	Square 2
V_4	-	Green Court 1
V_5	-	Tower 1 (front view)
V_6	-	Valley entrance
V_7	-	Green Court 2
V_8	-	Green Crescent
V_9	-	Square 5
V_{10}	-	Square 1
V_{11}	-	Park View
V_{12}	-	Lake View
V_{13}	-	Tower 2 (back view)
V_{14}	-	Tower 3 (Side view)
V_{15}	-	Physics building view
V_{16}	-	From South Towers
V_{17}	-	Over Square 4, with reflections
V_{18}	-	Through Information Centre, with reflections
V_{19}	-	Wivenhoe House & Playing fields
V_{20}	-	From inside room 6.314.

<u>**Visual sets**</u>

<u>**N-1 level**</u>

Set F (Facade strips)

F_1	=	Structural module 1	(large)
F_2	=	" " 2	(small)
F_3	=	Cladding	(metal & glass)
F_4	=	Structural module 2 + infill.	
F_5	=	Wall 1	(solid, concrete)
F_6	=	Wall 2	(finestrated, concrete,horizontal)
F_7	=	Wall 3	(" " vertical)
F_8	=	Structural module 2 + glass + infill	(concrete panel)
F_9	=	gateway 1	(one storey high)
F_{10}	=	gateway 2	(two storey high)
F_{11}	=	portico 1	(with wall + wooden doors)
F_{12}	=	portico 2	(with glass; transparent)
F_{13}	=	portico 3	(with glass + passageways)
F_{14}	=	landscape strip	
F_{15}	=	structural module 1 + infill.	(concrete panel)
F_{16}	=	Vertical glazed strip	(tower kitchens)
F_{17}	=	brick strip 1	(vertical, 1 window seam)
F_{18}	=	brick strip 2	(" 2 window seams)
F_{19}	=	brick strip 3	(" 3 " ")
F_{20}	=	brick strip 4	(" 5 " ")
F_{21}	=	LTB 1	(external stairwell column)
F_{22}	=	LTB 2	(back exit side)
F_{23}	=	LTB 3	(main-entrance side)
F_{24}	=	Hexagon Facade	
F_{25}	=	Library Facade	
F_{26}	=	Podium 1	(with fountain)
F_{27}	=	Podium 2	(with recessed pit)
F_{28}	=	Podium 3	(with raised tree-bed)
F_{29}	=	Roof Service-boxes	
F_{30}	=	Flight of steps	(monumental)
F_{31}	=	Waterscape facade	
F_{32}	=	Field	(slope, footpath, flat lawn)
F_{33}	=	Copse	
F_{34}	=	Stepped grass terrace	
F_{35}	=	Wivenhoe House Facade	
F_{36}	=	Platform undercarriage	(valley entrance)

N-2 level

Visual sets

Set D	Details	
D_1	Mullion	(smooth-faced, white, concrete, slender)
D_2	Beam end	(projecting)
D_3	Beam	
D_4	Size 1 beam	(Large, long)
D_5	" 2 "	(small, as in small, structural module)
D_6	" 3 "	(very small)
D_7	Polygonal	(LTB)
D_8	Glass Panel	
D_9	Size 1 glass panel	(very narrow, slit-like, Hexagon)
D_{10}	" 2 "	(narrow, Teach bldg)
D_{11}	" 3 "	(medium, Teach bldg)
D_{12}	" 4 "	(wide, ")
D_{13}	" 5 "	(Very wide, Library, Coffee Bar)
D_{14}	Slab	(Concrete)
D_{15}	Size 1 Slab	(thick, 2 or more storey high, Service Area Library)
D_{16}	" 2 "	(thin, wood-textured)
D_{17}	Concrete panel	(pebble faced, narrow, 2 storey high)
D_{18}	Glass door	
D_{19}	Wooden door	
D_{20}	Brick work	(structural)
D_{21}	pillar	(concrete, smooth faced, white)
D_{22}	wall piece	(concrete)
D_{23}	bright curtains	(in tower flats)
D_{24}	steps	(concrete, monumental)
D_{25}	Concrete	
D_{26}	glazed unit	(sports pavilion)
D_{27}	shaped-gable roof	(Wivenhoe House)
D_{28}	chimney stack, ornam.	
D_{29}	bay windows, stacked	
D_{30}	window surrounds	(stone)
D_{31}	extended corner	(Wivenhoe House)
D_{32}	Quoin	(")
D_{33}	Lantern	(")
D_{34}	Baroque door	(Wivenhoe House door)
D_{35}	Oriel window	(Wivenhoe House window)
D_{36}	Flat roof	
D_{37}	Paved footpath	
D_{38}	Slope	(smoothly lawned)
D_{39}	Grass	

D_{40}	Oak tree	
D_{41}	Birch tree	
D_{42}	Norwegian pine tree	
D_{43}	Sequoia tree	
D_{44}	cedar tree	
D_{45}	cork tree	
D_{46}	other tree	
D_{47}	lake	
D_{48}	white côlour	
D_{49}	pink colour	
D_{50}	black colour	
D_{51}	old red colour	
D_{52}	natural concrete colour	
D_{53}	hammered concrete finish	
D_{54}	smooth concrete finish	
D_{55}	pebbled concrete (panel)	
D_{56}	polygonally shaped	(window in LTB)
D_{57}	overhang	(Library)
D_{58}	long horizontal floor frame	(")
D_{59}	corner window	(LTB)
D_{60}	polygonal window stack	(")
D_{61}	concrete block infill	(")
D_{62}	metal mullion	
D_{63}	infill panel	(concrete)
D_{64}	window opening	(small, vertical)
D_{65}	floor piece	(tiled)
D_{66}	covered passageway	
D_{67}	metal window seam	(Tower window)
D_{68}	canopy	(entrance LTB)
D_{69}	platform	(concrete)
D_{70}	fountain	
D_{71}	recessed pit	(square 3)
D_{72}	raised tree bed	
D_{73}	geese	
D_{74}	flat lawn	
D_{75}	double sized	(gates)
D_{76}	vertical window seam 1	
D_{77}	" 2	
D_{78}	" 3	
D_{79}	" more than 3	

D_{80}	plaster finish	
D_{81}	monumental	
D_{82}	buttress	
D_{83}	cilindrical pillar	(2 or more storey high)
D_{84}	Road, access	

DATA FILE : SA v L

KY(X)

Value of q	Value of Q_q	Components
22	1	(SERVICES).
13	2	(SERVICES), (INDOOR GAMES).
9	3	(SERVICES), (INDOOR GAMES), (OUTDOOR GAMES).
8	2	(SERVICES, ACAD SOC, HOBBY CLUB, INDOOR GAMES, OUTDOOR GAMES), (POLIT CLUB).
5	1	(RETAIL, SERVICES, ACAD SOC, HOBBY CLUB, INDOOR GAMES, OUTDOOR GAMES, POLIT CLUB, SOCIAL CLUB).
4	1	(RETAIL, SERVICES, CATERING, ACAD SOC, ENTERTAINMENT, HOBBY CLUB, INDOOR GAMES, OUTDOOR GAMES, POLIT CLUB, SOCIAL CLUB, VOLUNTARY SERV).
3	1	(RETAIL, CATERING, SERVICES, HOUSING, ACAD SOC, ENTERTAINMENT, HOBBY CLUB, INDOOR GAMES, OUTDOOR GAMES, POLIT CLUB, SOCIAL CLUB, VOLUNTARY SERV).
2	2	(RETAIL, CATERING, SERVICES, ACAD SOC, HOBBY CLUB, HOUSING, ENTERTAINMENT, INDOOR GAMES, OUTDOOR GAMES, POLIT CLUB, SOCIAL CLUB, VOLUNTARY SERV), (STUDENT AGENCY).
1	1	(RETAIL, CATERING, SERVICES, ACAD SOC, HOBBY CLUB, INDOOR GAMES, STUDENT AGENCY, HOUSING, ENTERTAINMENT, OUTDOOR GAMES, POLIT CLUB, SOCIAL CLUB, VOLUNTARY SERV).
0	1	(ALL).

SA = Social Amenities   L = Locations

DATA FILE : SA v L

KX(Y)

Value of q	Value of Q_q	Components
9	1	(18).
7	1	(7, 18).
6	2	(7, 18, 9), (24).
5	3	(4), (7, 9, 16, 18, 30), (24).
4	5	(1), (2), (4, 18, 7, 9, 16, 24, 30), (11), (15).
3	3	(1, 7, 15, 16, 18, 24, 4, 9, 30, 6, 11), (2), (8, 12).
2	1	(1, 4, 7, 9, 11, 15, 16, 18, 24, 2, 30, 6, 8, 12).
1	1	(1, 2, 4, 7, 8, 9, 11, 12, 15, 16, 18, 24, 3, 14, 25, 26, 30, 19, 27, 28, 29, 6).
0	1	(1, 2, 3, 4, 7, 8, 9, 10, 11, 12, 14, 15, 16, 18, 20, 21, 24, 25, 26, 27, 28, 29, 30, 5, 6, 17, 19).

SA = Social Amenities L = Locations

DATA FILE : HOUSING (1972/3)

(N-1) v (N-2)

KY(X)

Value of q	Value of Q_q	Components
14	2	(S.H.-A), (S.H.-D).
13	3	(S.H.-A), (S.H.-C),(S.H.-D).
12	1	(S.H.-A), S.H.-C, S.H.-D).
11	2	(S.H.-A, S.H.-C, S.H.-D), (S.H.-B).
10	1	(S.H.-A, S.H.-C, S.H.-D, S.H.-B).
8	7	(CONTRACT-1), (CONTRACT-3), (CONTRACT-5), (CONTRACT-7), (WIV-HOUSE-A), (S.H.-A, S.H.-B, S.H.-C, S.H.-D), (WIV-HOUSE-B).
7	10	(CONTRACT-1, CONTRACT-5), (CONTRACT-3, CONTRACT-7), (HOUSE-CLACT-A, HOUSE-CLACT-B), (HOUSE-CLACT-C), (HOUSE-CLACT-D), (TOWERFLAT-B),(TOWERFLAT-STAFF-A), (WIV-HOUSE-A), (WIV-HOUSE-B), (S.H.-A, S.H.-B, S.H.-C, S.H.-D).
6	10	(CONTRACT-1, CONTRACT-3, CONTRACT-5, CONTRACT-7), (HOUSE-CLACT-A, HOUSE-CLACT-B, HOUSE-CLACT-C, HOUSE-CLACT-D), (HOUSE-COL-A), (HOUSE-COL-B), (TOWERFLAT-A), (TOWERFLAT-B), (GM6), (GMB, S.H.-B, S.H.-A, S.H.-C, S.H.-D, GM9, TOWERFLAT-STAFF-A), (TOWERFLAT-STAFF-B), (WIV-HOUSE-A, WIV-HOUSE-B).
5	7	(CONTRACT-1, CONTRACT-3, CONTRACT-5, CONTRACT-7, TOWERFLAT-A, TOWERFLAT-B), (HOUSE-CLACT-A, HOUSE-CLACT-B, HOUSE-CLACT-C, HOUSE-CLACT-D), (HOUSE-COL-A, HOUSE-COL-B), (GM1), (GM2, TOWERFLAT-STAFF-A, GM9, TOWERFLAT-STAFF-B, S.H.-A, S.H.-C, S.H.-D, GM8, S.H.-B), (GM6), (WIV-HOUSE-A, WIV-HOUSE-B).

KY(X) (Cont.) Data File : Housing (1972/3) (N-1) v (N-2)

Value of q	Value of Q_q	Components
4	5	(CONTRACT-1, CONTRACT-3, CONTRACT-5, CONTRACT-7, HOUSE-CLACT-A, HOUSE-CLACT-B, HOUSE-CLACT-C, HOUSE-CLACT-D, HOUSE-COL-A, TOWERFLAT-A, TOWERFLAT-B, HOUSE-COL-B, GM6), (GM1, GM2, TOWERFLAT-STAFF-A, TOWERFLAT-STAFF-B, GM9, S.H.-C, GM8, S.H.-A, S.H.-B, S.H.-D, WIV-HOUSE-A, WIV-HOUSE-B), (GM3), (GM4), (GM5).
3	2	(CONTRACT-1, CONTRACT-3, CONTRACT-5, CONTRACT-7, HOUSE-CLACT-A, HOUSE-CLACT-B, HOUSE-CLACT-C, HOUSE-CLACT-D, HOUSE-COL-A, HOUSE-COL-B, TOWERFLAT-A, TOWERFLAT-B, GM6), (GM1, GM2, GM5, GM8, GM9, TOWERFLAT-STAFF-A, TOWERFLAT-STAFF-B, S.H.-B, S.H.-C, GM4, S.H.-A, S.H.-D, WIV-HOUSE-A, GM3, WIV-HOUSE-B).
2	1	(CONTRACT-1, CONTRACT-3, CONTRACT-5, CONTRACT-7, HOUSE-CLACT-A, HOUSE-CLACT-B, HOUSE-CLACT-C, HOUSE-CLACT-D, HOUSE-COL-A, HOUSE-COL-B, TOWERFLAT-A, TOWERFLAT-B, GM6, WIV-HOUSE-A, WIV-HOUSE-B, S.H.-A, S.H.-B, S.H.-C, S.H.-D, GM1, GM2, GM8, GM9, TOWERFLAT-STAFF-A, TOWERFLAT-STAFF-B, GM4, GM3, GM5).
0	1	(ALL).

KX(Y)

Value of q	Value of Q_q	Components
16	1	(HDS).
15	2	(DX), (HDS).
14	3	(DX), (B,T),(HDS).
13	2	(DX, HDS), (B), (T).
12	3	(DX, HDS, TX), (B, T), (PI).
10	2	(DX, HDS, BX, TX, PI), (B, T, BRD).
9	2	(DX, TX, HDS, PI, BX), (H,B,T, BRD).
8	3	(LRX, HDS, DX, BX, TX, PI), (H, B, T, BRD), (PHM).
7	3	(KX, BX, TX, HDS, DX, PI, LRX), (LS, B, T, BRD, H), (PHM).
6	2	(KW, B, T, LS, H, PHM, BRD), (KX, BX, TX, HDS, DX, PI, LRX).
5	1	(KW, B, T, BRD, DX, LS, H, PM, PHM, KX, LS, BX, TX, HDS, BF, PI, LI).
3	1	(KW, D, LS, H, B, T, PM, BRD, PHM, LRX, DX, HDS, BR, LA, KD, KX, BX, TX, PI, BF, ED, LI, SB1)
2	1	(KW, LRX, D, DX, LS, H, B, T, HDS, PM, PHM, BR, BRD, LA, KX, BX, TX, PI, U, BF, ED, LI, KD, SB1, KDX).
1	1	(KW, LRX, D, DX, LS, S, H, B, T, HDS, PM, PHM, BR, BRD, U, LA, KX, BX, TX, PI, KDX, BL, DS, BF, ED, LI, SB1, KD, KA, BS4D).
0	1	(ALL)

DATA FILE : HOUSING (1973/4) (N-1) v (N-2)

KY(X)

Value of q	Value of Q_q	Components
14	2	(S.H.-A), (S.H.-D).
13	3	(S.H.-A), (S.H.-C), (S.H.-D).
12	1	(S.H.-A, S.H.-C, S.H.-D).
11	2	(S.H.-A, S.H.-C, S.H.-D), (S.H.-B).
10	1	(S.H.-A, S.H.-C, S.H.-D, S.H.-B).
8	7	(CONTRACT-1),(CONTRACT-3), (CONTRACT-5), (CONTACT-7), (WIV-HOUSE-A), (S.H.-A, S.H.-B, S.H.-C, S.H.-D), (WIV-HOUSE-B).
7	15	(CONTACT-1, CONTRACT-5), (CONTRACT-3, CONTRACT-7), (HOUSE-CLACT-A), (HOUSE-CLACT-B), (HOUSE-CLACT-C), (HOUSE-CLACT-D), (WEST-END-A), (WEST-END-B), (WEST-END-C), (WEST-END-D), (WEST-END-E), (TOWERFLAT-B), (TOWERFLAT-STAFF-A), (WIV-HOUSE-A, WIV-HOUSE-B), (S.H.-A, S.H.-B, S.H.-C, S.H.-D).
6	12	(CONTRACT-1, CONTRACT-3, CONTRACT-5, CONTRACT-7, HOUSE-CLACT-A, HOUSE-CLACT-B, HOUSE-CLACT-C, HOUSE-CLACT-D), (HOUSE-COL-A), (HOUSE-COL-B), (WEST-END-A, WEST-END-B, WEST-END-C, WEST-END-D, WEST-END-E), (TOWERFLAT-A, AVON WAY), (TOWERFLAT-B), (GM6), (GM7), (GM8, S.H.-B, S.H.-A, S.H.-C, S.H.-D, GM9, TOWERFLAT-STAFF-A), (TOWERFLAT-STAFF-B), (WIV-HOUSE-A, WIV-HOUSE-B).
5	8	(CONTRACT-1, CONTRACT-3, CONTRACT-5, CONTRACT-7, TOWERFLAT-A, TOWERFLAT-B, AVON WAY), (HOUSE-CLACT-A, HOUSE-CLACT-B, HOUSE-CLACT-C, HOUSE-CLACT-D), (HOUSE-COL-A, HOUSE-COL-B), (WEST-END-A, WEST-END-B, WEST-END-C, WEST-END-D, WEST-END-E), (GM1), (GM2,

KY(X) (cont).

Value of q	Value of Q_q	Components
5 cont.	8	TOWERFLAT-STAFF-A, GM9, TOWERFLAT-STAFF-B, S.H.-A, S.H.-C, S.H.-D, GM8, S.H.-B), (GM6, GM7), (WIV-HOUSE-A, WIV-HOUSE-B).
4	5	(CONTRACT-1, CONTRACT-3, CONTRACT-5, CONTRACT-7, HOUSE-CLACT-A, HOUSE-CLACT-B, HOUSE-CLACT-C, HOUSE-CLACT-D, HOUSE-COL-A, TOWERFLAT-A, TOWERFLAT-B, AVON WAY, HOUSE-COL-B,WEST-END-A, WEST-END-B, WEST-END-C, WEST-END-D, WEST-END-E, GM6, GM7), (GM1, GM2, TOWERFLAT-STAFF-A, TOWERFLAT-STAFF-B, GM9, S.H.-C, GM8, S.H.-A, S.H.-B, S.H.-D, WIV-HOUSE-A, WIV-HOUSE-B),(GM3), (GM4), (GM5).
3	1	(CONTRACT-1, CONTRACT-3, CONTRACT-5, CONTRACT-7, HOUSE-CLACT-A, HOUSE-CLACT-B, HOUSE-CLACT-C, HOUSE-CLACT-D, HOUSE-COL-A, HOUSE-COL-B, WEST-END-A, TOWERFLAT-A, TOWERFLAT-B, AVON WAY, GM6, GM7, WEST-END-B, WEST-END-C, WEST-END-D, WEST-END-E, S.H.-A, S.H.-B, S.H.-C, S.H.-D, GM2, GM8, GM9, TOWERFLAT-STAFF-A, TOWERFLAT-STAFF-B, WIV-HOUSE-A, WIV-HOUSE-B, GM1, GM4, GM3, GM5).
0	1	(ALL).

DATA FILE : HOUSING (1973/4) (N-1) v (N-2)

KX(Y)

Value of q	Value of Q_q	Components
20	1	(DX).
19	2	(DX), (TX).
18	3	(DX), (TX), (HDS).
17	3	(DX), (BX, TX), (HDS).
14	3	(KX, BX, TX, DX, HDS), (B, T), (PI).
13	4	(KX, BX, TX, DX, HDS), (LRX), (B, T), (PI).
12	2	(KX, BX, TX, DX, HDS, PI, LRX), (B,T).
10	3	(KX, DX, LX, TX, LRX, HDS, PI), (B, T, BRD), (PM).
9	3	(KX, DX, BX, TX, HDS, LRX, PI), (H, B, T, BRD), (PM).
8	4	(KX, LRX, DX, BX, TX, HDS, PM, LA, PI), (LS), (H, B, T, BRD), (PHM).
7	3	(KX, LRX, DX, BX, TX, HDS, PI, PM, LA), (LS, B, T, BRD, H), (PHM).
6	2	(KN, B, T, LS, H, PHM, BRD), (KX, LRX, DX, BX, TX, HDS, PI, PM, LA, LI).
5	1	(KW, B, T, BRD, DX, LS, H, PM, PHM, KX, LRX, BX, TX, HDS, BF, PI, LA, SB1, LI).
3	1	(KW, D, LS, H, B, T, PM, BRD, PHM, LRX, DX, HDS, BR, LA, KD, KX, BX, TX, PI, BF, ED, LI, SB1).
2	1	(KW, LRX, D, DX, LS, H, B, T, HDS, PM, PHM, BR, BRD, LA, KX, BX, TX, PI, U, BF, EV, LI, SB1, KD, BS3D, KDX).
1	1	(KW, LRX, D, DX, LS, S, H, B, T, HDS, PM, PHM, BR, BRD, U, LA, KX, BX, TX, PI, SB1, KDX, BL, DS, BF, ED, LI, SB2, KD, KA, BS4D, BS3D).
0	1	(ALL).

DATA FILE : SvL

KY(X)

Value of q	Value of Q_q	Components
12	1	(GOOD WEATHER AREA).
5	2	(PARKING), (GOOD WEATHER AREA).
4	3	(PARKING), (THEATRE), (GOOD WEATHER AREA).
3	4	(PARKING), (DANCEFLOOR), (THEATRE), (GOOD WEATHER AREA).
2	7	(PARKING, GOOD WEATHER AREA, THEATRE), (T.V. ROOM), (DANCEFLOOR, CONCERT), (COMMON ROOM), (STUDENTS COUNCIL), (SECURITY), (EXHIBITION).
1	3	(PARKING, THEATRE, GOOD WEATHER AREA, DANCEFLOOR, CONCERT, SECURITY, EXHIBITION, STUDENTS COUNCIL, COMMON ROOM), (T.V. ROOM), (CAREERS).
0	1	(LAUNDERETTE, BANK, POST OFFICE, T.V. ROOM, GOOD WEATHER AREA, HOUSING OFFICE, CAREERS, PARKING, NURSERY, DANCEFLOOR, COMMON ROOM, LIBRARY, HEALTH SERVICE, STUDENTS COUNCIL, SECURITY, THEATRE, CONCERT, EXHIBITION, NIGHTLINE, GAMES ROOM, TOWERS OFFICE).

KX(Y)

Value of q	Value of Q_q	Components
4	3	(8), (18), (24).
3	7	(1), (7), (8), (12), (18), (24), (30).
2	8	(1), (2), (7, 18), (8), (12), (15), (24), (30).
1	5	(1, 7, 21, 24, 2, 18, 9, 12, 28, 30, 14, 15), (4), (8), (10), (26).
0	1	(1, 7, 8, 9, 12, 14, 16, 18, 21, 24, 25, 26, 27, 28, 29, 30, 2, 15, 4, 10, 11, 3, 20).

S = Services L = Locations

Data File : SOC v L

KY(X)

Value of q	Value of Q_q	Components
3	5	(CROSS COUNTRY), (SQUASH), (CHESS CLUB), (GAY LIB), (MUSIC).
2	8	(CROSS COUNTRY), (GOLF), (SQUASH), (CHESS CLUB), (CHRISTIAN UNION), (GAY LIB), (MUSIC), (T.A.S.),
1	11	(CROSS COUNTRY), (DARTS), (GOLF), MOUNT-ING, GAY LIB, CHRISTIAN UNION, LAT AM SOC, MUSIC, SOC SOC, SPART LEAGUE, WOMENS LIB, METH SOC, T.A.S.), (SAILING), (SQUASH), (AM RADIO), (CHESS CLUB), (IS & JEW), (PHOTO SOC), (ENTS, ATHLETICS UNION).
0	3	(ATH, CRICKET, CROSS COUNTRY, FOOTBALL, GOLF, HOCKEY, RUGBY, IS & JEW, NIGHTLINE, FENCING, SAILING, SQUASH, AM RADIO, FOLK DANCE, GAY LIB, MUSIC, RAILWAY SOC, WOMENS LIB, KEEP FIT, ENTS, ATHLETICS UNION, MOUNT-ING, VOLLEYBALL, ANTI-BLOOD SP, ART-SOC, ASIAN SOC, ASTRON, CHEM-SOC, CHRISTIAN UNION, COMMUNIST, CONSERVATION GRP, DEBATING SOC, ECON SOC, FILM SOC, FOLK CLUB, INT SOC, LABOUR CLUB, LAT AM SOC, SOC SOC, SPART LEAGUE, T.A.S., 3RD WORLD, YOGA, DISCUSSION GRP, SWIM, METH SOC, PRIS ACTION, DARTS, SNOOKER, TABLE TEN, PHOTO SOC, CROQUET, AD PLAYDRP, BRIDGE, CHESS CLUB, ELECTRONICS SOC, IVS, R.C. GROUP, URE, UTO, WYVERN, CHINESE SOC, MIME GROUP), (BAD, BASK, KARATE, NETBALL), (NUS TRAVEL).

Soc = Societies L = Locations

DATA FILE : SOC v L

KX(Y)

Value of q	Value of Q_q	Components
27	1	(7).
11	2	(7), (9).
9	3	(7), (9), (18).
8	4	(7), (9), (18), (30).
4	6	(6), (7, 9), (12), (15), (18), (30).
3	9	(4), (6), (7, 9), (12), (15), (18), (19), (24), (30).
2	9	(4), (6), (7, 9, 18), (12), (15), (16), (19), (24), (30).
1	7	(1, 7, 9, 18, 24, 6, 12), (4), (11), (15), (16), (19), (30).
0	3	(1, 7, 11, 16, 18, 4, 6, 9, 24, 3, 5, 12, 14, 15, 17, 27, 28, 29, 2), (8), (19).

Soc = Societies L = Locations

DATA FILE : G v D (1970/1)

KY(X)

Value of q	Value of Q_q	Components
23	1	(CLCS).
20	2	(CLSS), (CLCS).
17	1	(CLSS, CLCS).
16	2	(CLPS), (CLSS, CLCS).
14	1	(CLPS, CLCS, CLSS).
9	2	(CLPS, CLSS, CLCS), (CLJS).
8	1	(CLPS, CLSS, CLCS, CLJS).
7	2	(MRAS), (MRAA, CLCS, CLPS, CLSS, CLJS).
5	3	(MRGA, CLCS, MRAS, MRAA, CLPS, CLSS, CLJS), (TECO), (TSNO).
4	1	(MRGA, CLCS, MRAS, MRAA, CLPS, CLSS, CLJS, TCHA, TECO, TSNO).
3	1	(MRGS, CLSS, CLCS, CRGA, MRAS, MRAA, CLPS, CLJS, TCHA, TECO, TSNO).
1	3	(MRGS, MRAA, CLPS, CLSS, CLCS, CLJS, MRGA, MRAS, TCHA, TECO, TSNO, SNCH), (CLPR2), (LSLA).
0	2	(MRGS, MRGA, MRAS, MRAA, CLPS, CLSS, CLCS, CLJS, LSLA, LLAA, LJLA, SNMN, CLPR1, TCHA, TECO, TSNO, SNCH, CLPR2), (SNMP).

KX(Y)

Value of q	Value of Q_q	Components
6	6	(AC), (CC), (EE), (FI), (LB), (MN).
5	5	(AC, EP), (CC, EE, PH, CH), (FI), (LB), (MN).
4	4	(AC, EP, FI, MN), (CC, CH, EE, LA, PH), (HX), (LB).
3	1	(AC, CC, EP, FI, MN, SP, CH, EE, LA, PH, DS, EC, LB, MA, PC, HX).
2	1	(AC, CA, CC, CS, DS, EC, EE, EP, FI, HX, LA, LT, MA, MN, PC, PH, SC, SP, CH, LB, GV), (AR), (MS), (WH).
1	1	(AC, CA, CC, CH, CS, DS, EC, EE, EP, FI, GV, HG, HX, LA, LB, LT, MA, MN, MS, PC, PH, PS, SC, SP, SS, ST, WH, AR, DB).
0	2	(AC, AR, CA, CC, CH, CS, DS, EC, EE, EP, FI, GV, HG, HX, LA, LB, LT, MA, MN, MS, PC, PH, PS, SC, SP, SS, ST, WH, DB, ET), (ML).

G = Staff Grades D = Departments

DATA FILE : G v D (1971/2)

KY(X)

Value of q	Value of Q_q	Components
22	1	(CLCS),
21	2	(CLSS), (CLCS).
16	2	(CLPS), (CLSS, CLCS).
14	1	(CLPS, CLCS, CLSS).
8	2	(MRAS), (CLPS, CLSS, CLCS, CLJS).
7	1	(MRAS, CLSS, CLPS, CLCS, CLJS).
6	2	(MRAS, CLSS, CLCS, CLPS, CLJS), (TSNO).
5	1	(MRGA, CLSS, MRAS, CLPS, CLCS, CLJS, TSNO, MRAA).
4	1	(MRGA, CLSS, CLCS, MRAS, MRAA, CLPS, CLJS, TSNO, MRAJ, TCHA, TECO).
3	1	(MRGS, CLSS, CLCS, MRGA, MRAS, MRAJ, MRAA, CLPS, CLJS, TCHA, TECO, TSNO).
2	1	(MRGS, CLPS, CLSS, CLCS, MRGA, MRAS, MRAJ, MRAA, CLPR2, CLJS, TCHA, TECO, TSNO).
1	1	(MRGS, MRAS, MRAA, CLPS, CLSS, CLCS, CLJS, MRGA, MRAJ, CLPR2, SNMN, SNCH, LSLA, TCHA, TECO, TSNO, CSNO, CECO, CJRO).
0	2	(MRGS, MRGA, MRAS, MRAJ, MRAA, CLPR2, CLPR1, CLPS, CLSS, CLCS, CLJS, TCHA, TECO, TSNO, CCHA, CSNO, CECO, CJRO, SNMN, SNCW, SNCH, SNSP, LSLA, LLAA, LJLA), (SNMD).

G = Staff Grades   D = Departments

DATA FILE : G v D (1971/2)

KX(Y)

Value of q	Value of Q_q	Components
10	1	(CC).
9	2	(CC), (MN).
8	4	(AC), (CC), (FI), (MN).
7	3	(AC, MN), (CC), (FI).
6	2	(AC, FI, MN), (CC).
5	6	(AC, FI, MN), (CC, EE), (EP), (LA), (LB), (SP).
4	3	(AC, EP, FI, HG, MN, SP), (CC, CH, EE, LA, PH), (LB).
3	3	(AC, CA, EC, EP, FI, GV, HG, MN, PC, SC, SP, CC, CH, DB, EE, LA, MA, PH), (HX), (LB).
2	3	(AC, CA, CC, EC, EE, EP, FI, GV, HG, HX, LA, LB, LT, MA, MN, PC, PS, SC, SP, CH, DB, DS, PH), (AR), (WH).
1	1	(AC, CA, CC, CH, CS, DS, EC, EE, EP, FI, GV, HG, HX, LA, LB, LT, MA, MN, MS, PC, PH, PS, SC, SP, SS, ST, WH, AR, DB).
0	2	(AC, AR, CA, CC, CH, CS, DB, DS, EC, EE, EP, FI, GV, HG, HX, LA, LB, LT, MA, MN, MS, PC, PH, PS, SC, SP, SS, ST, WH), (ML).

G = Staff Grades   D = Departments

DATA FILE : G v D (1973/4)

KY(X)

Value of q	Value of Q_q	Components
25	1	(CLSS).
19	2	(CLSS), (CLCS).
19	1	(CLSS, CLCS).
17	2	(CLPS), (CLSS, CLCS).
14	1	(CLPS, CLCS, CLSS).
9	2	(MRAS), (CLPS, CLSS, CLCS, CLJS).
8	2	(MRGS, CLCS, CLPS, CLSS, CLJS), (MRAS).
7	1	(MRGS, CLCS, MRAS, MRAJ, CLPS, CLSS, CLJS).
6	2	(MRGS, CLSS, CLCS, MRAS, CLPS, CLJS, MRAJ), (TSNO).
5	1	(MRGS, MRAS, CLPS, CLSS, CLCS, CLJS, MRAJ, TECO, TSNO, MRGA).
4	1	(MRGS, MRAS, CLPS, CLSS, CLCS, CLJS, MRAJ, TSNO, MRGA, CLPR2, TECO).
3	1	(MRGS, MRAS, MRAJ, CLPS, CLSS, CLCS, CLJS, SNMN, MRGA, CLPR2, TECO, TSNO, TCHA, MRAA).
1	2	(MRGS, MRGA, MRAS, MRAJ, MRAA, CLPS, CLSS, CLCS, CLJS, CNMN, SNCH, SNSP, CLPR2, TCHA, TECO, TSNO, CSNO, CECO), (LSLA).
0	1	(MRGS, MRGA, MRAS, MRAJ, MRAA, CLPR2, CLPS, CLSS, CLCS, CLJS, LSLA, LLAA, LJLA, TECO, TSNO, CCHA, CSNO, CECO, CJRO, SNMN, SNCH, SNSP, SNCW, TCHA, SNMD).

G = Staff Grades   D = Departments

DATA FILE G v D (1973/4)

KX(Y)

Value of q	Value of Q_q	Components
11	1	(CC).
9	2	(CC), (MN).
8	3	(CC), (FI), (MN).
7	5	(AC), (CC), (FI), (LB), (MN).
6	6	(AC, FI, MN), (CC), (EP), (HX), (LA), (LB).
5	5	(AC, CC, EP, FI, MN), (EE), (HX, WH), (LA), (LB).
4	3	(AC, CC, EP, FI, HX, LB, MN, EE, LA, GV, PC, WH, SP, CH, PH), (HG), (SS).
3	1	(AC, CA, CC, EC, EP, FI, GV, HG, HX, LA, LB, MN, PC, SC, SS, WH, CH, EE, MA, PH, SP).
2	2	(AC, CA, CC, DS, EC, EE, EP, FI, GV, HG, HX, LA, LB, MA, MN, MS, PC, SC, SP, SS, WH, CH, DB, ET, PH, LT), (AR).
1	2	(AC, CA, CC, CH, CS, DS, EC, EE, EP, FI, GV, HG, HX, LA, LB, LT, MA, MN, MS, PC, PH, PS, SC, SP, SS, WH, AR, DB, ET), (ML).
0	1	(AC, AR, CA, CC, CH, CS, DB, DS, EC, EE, EP, ET, FI, GV, HG, HR, HX, LA, LB, LT, MA, ML, MN, MS, PC, PH, PS, SC, SP, SS, ST, WH).

G = Staff Grades D = Departments

DATA FILE : AG v D (1970/1)

KY(X)

Value of q	Value of Q_q	Components
15	1	(ACPR).
10	1	(ACPR, ACLC).
9	1	(ACPR, ACLS, ACLC, EXPA).
7	1	(ACPR, ACLS, ACLC, EXPO, EXPA).
6	1	(ACPR, ACLS, ACLC, EXPO, ECPF, EXPA).
4	2	(ACPR, ACLS, ACLC, EXPO, ECPF, EXPA), (ACRD).
3	2	(ACPR, ACRD, ACLS, ACLC, EXPO, ECPF, EXPA), (EXPS).
2	1	(ACPR, ACRD, ACLS, ACLC, EXPS, EXPO, ECPF, EXPA).

KX(Y)

7	1	(CC, CH).
6	1	(CC, CH, MA).
5	1	(CC, CH, MA, PH, SC).
4	1	(CC, CH, EC, EE, LA, LT, MA, PH, SC).
3	1	(AR, CC, CH, PH, EC, EE, GV, LA, LT, MA, SC).
1	1	(AR, CC, CH, EC, EE, GV, LA, LT, MA, PH, SC, EP, HR).
0	1	(AC, AR, CC, CH, DB, EC, EE, EP, FI, GV, LA, LB, LT, MA, PH, SC, HR).

DATA FILE : AG v D (1971/2)

KY(X)

Value of q	Value of Q_q	Components
15	1	(ACPR).
10	1	(ACPR, ACLC).
8	1	(ACPR, ACLC, EXPA).
7	1	(ACPR, ACLS, ACLC, EXPA).
5	1	(ACPR, ACRD, ACLS, ACLC, EXPO, EXPF, EXPA).
4	1	(ACPR, ACRD, ACLS, ACLC, EXPS, EXPO, EXPF, EXPA).
1	1	(ACPR, ACRD, ACLS, ACLC, EXPO, EXPS, EXPO, EXPF, EXPA).

KX(Y)

7	2	(LA), (PH).
6	2	(CC), (LA, PH).
5	1	(CC, LA, LT, PH, SC, CH).
4	1	(CC, CH, LA, LT, PH, SC, GV, EE).
3	1	(CC, CH, EE, GV, LA, LT, PH, SC, EC, MA).
2	1	(AR, CC, LA, LT, PH, SC, CH, DB, EC, EE, GV, MA).
0	1	(AC, AR, CC, CH, DB, EC, EE, EP, FI, GV, LA, LB, LT, MA, PH, SC).

DATA FILE : AG v D (1972/3)

KY(X)

Value of q	Value of Q_q	Components
14	1	(ACPR).
10	1	(ACPR, ACLC).
8	2	(ACPR, ACLS, ACLC, EXPA), (ACRD).
6	2	(ACPR, ACRD, ACLS, ACLC, EXPO, EXPF, EXPA), (EXPS).
5	1	(ACPR, ACRD, ACLS, ACLC, EXPS, EXPO, EXPF, EXPA).
1	1	(ACPR, ACRD, ACLS, ACLC, EXPC, EXPS, EXPO, EXPF, EXPA).

KX(Y)

7	2	(EE), (PH).
6	2	(CC, EE, GV, MA, PH), (LA).
5	1	(CC, EE, GV, MA, PH, EC, LA, SC).
4	1	(CC, EC, EE, GV, LA, MA, PH, SC, CH).
3	1	(CC, CH, EC, EE, GV, LA, MA, PH, SC, LT).
2	1	(AR, CC, EE, GV, MA, PH, CH, EC, LA, LT, SC).
1	1	(AR, CC, CH, EC, EE, GV, LA, LB, LT, MA, PH, SC, DB).
0	1	(AC, AR, CC, CH, EC, EE, EP, FI, GV, LA, LB, LT, MA, PH, SC, DB, HR).

DATA FILE : AG v D (1973/4)

KY(X)

Value of q	Value of Q_q	Components
14	1	(ACPR).
12	2	(ACPR), (ACLC).
10	1	(ACPR, ACLC, ACLS).
9	1	(ACPR, ACRD, ACLS, ACLC).
8	2	(ACPR, ACRD, ACLS, ACLC), (EXPO).
7	1	(ACPR, ACRD, ACLS, ACLC, EXPO, EXPF, EXPA).
6	1	(ACPR, ACRD, ACLS, ACLC, EXPO, EXPF, EXPA, EXPS).
1	1	(ACPR, ACRD, ACLS, ACLC, EXPC, EXPS, EXPO, EXPF, EXPA).

KX(Y)

7	2	(CC, EE), (PH).
6	2	(CC, EE, GV, MA, PH, SC), (LA).
5	1	(CC, CH, EC, EE, GV, LA, MA, PH, SC).
4	1	(CC, CH, EC, EE, GV, LA, LT, MA, PH, SC).
3	1	(CC, CH, DB, EC, EE, GV, LA, LT, MA, PH, SC).
2	1	(AR, CC, CH, EE, GV, LA, LT, MA, PH, SC, DB, EC).
1	1	(AR, CC, CH, EC, EE, GV, LA, LT, MA, PH, SC, DB, HR, LB).
0	1	(AC, AR, CC, CH, EC, EE, EP, FI, GV, LA, LB, LT, MA, PH, SC, DB, HR).

DATA FILE : DP v DS (1970/1)

KY(X)

Value of q	Value of Q_q	Components
4	1	(MATHS).
2	6	(GOV), (SOC), (ECON), (MATHS), (PHYS), (CHEM).
0	8	(ART), (LIT), (LANG), (GOV, ECON, MATHS, SOC, PHYS, CHEM), (CC), (EE).

KX(Y)

Value of q	Value of Q_q	Components
1	5	(ECON/GOV), (MATHS/SOC), (MATHS/ECON), (THEO/PHYS), (CHEM-PHYS).
0	6	(ART), (LIT), (LANG), (GOV(CS)), (GOV(SS)), (ECON/GOV, ECON, MATHS/ECON, MATHS/SOC, MATHS/POL, MATHS, THEO/PHYS, SOC(CS), SOC(SS), PHYS, CHEM-PHYS, BIOL-CHEM, CHEM).

DATA FILE : DP v DS (1971/2)

KY(X)

Value of q	Value of Q_q	Components
4	3	(GOV), (SOC), (MATHS).
3	4	(GOV), (SOC), (MATHS), (PHYS).
2	6	(GOV), (SOC), (ECON), (MATHS), (PHYS), (CHEM).
1	5	(GOV, SOC), (ECON), (MATHS), (PHYS), (CHEM).
0	6	(ART), (LIT), (LANG), (GOV, SOC, ECON, MATHS, PHYS, CHEM), (CC), (EE).

KX(Y)

1	6	(ECON/GOV), (MATHS/SOC), (MATHS/ECON), (THEOPHYS), (CHEM-PHYS), (GOV/SOC(CS), GOV/SOC(SS)).
0	6	(ART), (LIT), (LANG), (GOV(CS), GOV(SS), ECON/GOV, GOV/SOC(CS), GOV/SOC(SS), ECON, MATHS/ECON, SOC(CS), SOC(SS), MATHS/SOC, MATHS/POL, MATHS, THEOPHYS, PHYS, CHEM-PHYS, APP-PHYS, BIOL-CHEM, CHEM), (CPTG-SC), (EE).

DP = Departments DS = Degree Schemes

DATA FILE : DP v DS (1972/3)

KY(X)

Value of q	Value of Q_q	Components
5	2	(GOV), (SOC).
4	3	(GOV), (SOC), (MATHS).
3	4	(GOV), (SOC), (MATHS), (PHYS).
2	8	(GOV), (SOC), (ECON), (MATHS), (PHYS), (CHEM), (EE), (HIST).
1	8	(LIT), (GOV,SOC), ECON), (MATHS), (PHYS), (CHEM), (EE), (HIST).
0	5	(ART), (LIT, HIST, GOV, SOC, ECON, MATHS, PHYS, CHEM), (LANG), (CC), (EE).

KX(Y)

1	9	(ECON/GOV), (MATHS/SOC), (MATHS/ECON), (THEOPHYS), (CHEM-PHYS), (GOV/SOC(CS), GOV/SOC(SS)), (HIST/GOV), (HIST/LIT), (HIST/SOC).
0	6	(ART), (LIT, HIST/LIT, HIST/GOV, HIST/SOC, GOV(CS), GOV(SS)), (ECON/GOV, GOV/SOC(CS), GOV/SOC(SS), SOC(CS), SOC(SS), MATHS/SOC, ECON, MATHS/ECON, MATHS/POL, MATHS, THEOPHYS, PHYS, CHEM-PHYS, APP-PHYS, BIOL-CHEM, CHEM).

DATA FILE : DP v DS (1973/4)

KY(X)

Value of q	Value of Q_q	Components
7	1	(MATHS).
6	2	(GOV), (MATHS).
5	3	(GOV), (SOC), (MATHS).
3	6	(GOV), (SOC), (CC), (MATHS, (PHYS),(CHEM).
2	9	(GOV), (SOC), (ECON), (CC), (MATHS), (PHYS), (CHEM), (EE), (HIST).
1	9	(LIT), (GOV, SOC), (ECON), (CC), (MATHS), (PHYS), (CHEM), (EE), (HIST).
0	4	(ART), (LIT, HIST, GOV, SOC, ECON, MATHS, CC, PHYS, CHEM), (LANG), (EE).

DP = Departments   DS = Degree Schemes

DATA FILE : DP v DS (1973/4)

KX(Y) Value of q	Value of Q_q	Components
1	12	(ECON/GOV), (MATHS/SOC), (MATHS/ECON), THE (THEO/PHYS), (CHEM/PHYS), (GOV/SOC(CS), GOV/SOC(SS)), (HIST/GOV), (HIST/LIT), (HIST/SOC), (GOV/STATS), (MATH-CPTG), (MATH-CHEM).
0	4	(ART), (LIT, HIST/LIT, HIST/GOV, HIST/SOC, GOV(CS), (GOV(SS), ECON/GOV, GOV/SOC(CS), GOV/SOC(SS), GOV/STATS, SOC(CS), SOC(SS), MATHS/SOC, ECON, MATHS/ECON, MATHS/POL, MATHS, THEOPHYS, MATH-CPTG, MATH-CHEM, PHYS, CHEM-PHYS, APP-PHYS, CPTG-SC, CPTERS-CPTG, CPTER-SYS, BIOL-CHEM, CHEM), (LANG), (EE, CPTG/COMM-ENG, TELECOMM-ENG).

DP = Departments   DS = Degree Schemes

DATA FILE : DP v DS (1974/5)

KY(X)

Value of q	Value of Q_q	Components
9	1	(SOC).
8	3	(GOV), (SOC), (MATHS).
4	7	(LANG), (GOV), (SOC), (CC), (MATHS), (PHYS), (CHEM).
3	10	(LIT), (LANG, SOC), (GOV), (ECON), (CC), (MATHS), (PHYS), (CHEM), (EE), (HIST).
1	9	(ART), (LIT), (LANG, SOC, GOV), (ECON), (CC, MATHS), (PHYS, CHEM), (EE), (HIST), (PHIL).
0	3	(ART, LIT, HIST; GOV, SOC, LANG, ECON, MATHS, PHIL, CC, PHYS, CHEM), (EE), (BIOL).

KX(Y)

1	17	(ECON/GOV), (MATHS/SOC), (MATHS/ECON), (THEOPHYS), (CHEM-PHYS, MAT-SC), (GOV/SOC(CS), GOV/SOC(SS)), (HIST/GOV), (HIST/LIT), (HIST/SOC), (GOV/STATS), (MATH-CPTG, MATHS/COMPTG), (MATH-CHEM), (LING/SOC, LANG/SOC, LING/SOC(SS)), (LIT/ART), (LANG/GOV), (PHIL/GOV), (PHIL/SOC).
0	3	(ART, LIT/ART, LIT, HIST/LIT, HIST/GOV, HIST/SOC, GOV(CS), GOV(SS), ECON/GOV, GOV/SOC(CS), (GOV/SOC(SS), GOV/STATS, LANG/GOV,PHIL/GOV, SOC(CS), (SOC(SS), MATHS/SOC, LING/SOC, LANG/SOC, LING/SOC(SS), PHIL/SOC, ECON, MATHS/ECON, MATHS/POL, MATHS, THEOPHYS, MATH-CPTG, MATH-CHEM, MATHS-COMPTG, LANG, PHYS, CHEM-PHYS, APP-PHYS, MAT-SC, CPTG-SC, CPTERS-CPTG, CPTER-SYS, BIOL-CHEM, CHEM), (EE, CPTG/COMM-ENG, TELECOMM-ENG), (BIOL-SC).

DP = Departments DS = Degree Schemes

DATA FILE : P v C (1972/3)

KY(X)

Value of q	Value of Q_q	Components
19	1	(1).
9	3	(1), (22H), (28L).
8	2	(1, 22H), (28L).
7	3	(1, 22H, 65H), (23M), (28L).
6	3	(1, 22H, 23M, 28L, 33P, 51S, 65M, 8E), (9F), (53P).
5	6	(1, 8E, 9E, 22H, 23M, 28L, 33P, 51S, 53P, 65M), (3E), (13C), (19EE), (68LA), (71C).
4	4	(1, 3E, 8E, 13C, 22H, 23M, 28L, 33P, 51S, 53P, 65M, 71C, 76LAY, 68LA). (19EE), (20EE), (72EE).
3	5	(1, 3E, 5G, 8E, 9E, 13C, 19EE, 20EF, 22H, 23M, 28L, 33P, 51S, 53P, 64A, 65M, 67S, 68LA, 71C, 73LA, 76LAY, 32EE), (12G), (24M), (45P), (72EE).
2	9	(1, 2M, 3E, 5G, 8E, 9E, 10M, 13C, 17L, 19EE, 20EE, 22H, 23M, 24M, 28L, 32EE, 33P, 37M, 39L, 45P, 51S, 53P, 58C, 64A, 65M, 66P, 67S, 68LA, 71C, 72EE, 73LA, 76LAY, 77LAY, 90LAY, 121G), (4G), (12G), (42L), (61LA), (107ST), (112ST), (129ST), (136ST),
1	7	(1, 2M, 3E, 4G, 5G, 6M, 8E, 9E, 10M, 12G, 13C, 14L, 15G, 17L, 19EE, 20EE, 22H, 23M, 24M, 28L, 32EE, 33P, 34C, 37M, 39L, 42L, 45P, 47G, 51S, 52LIB, 53P, 58C, 59C, 61LA, 64A, 65M, 66P, 67S, 68LA, 70L, 71C, 72EE, 73LA, 76LAY, 77LAY, 78LAY, 79LAY, 80LAY, 82LAY, 83LAY, 85LAY, 86LAY, 87LAY, 89LAY, 90LAY, 91LAY, 112ST, 121G, 27LA, 44E, 62G, 35C, 107ST, 129ST, 136ST, 109ST, 30C, 144ST, 41C), (7S), (16G), (55H, 151G), (119C), (127LIB), (131M, 132P).
0	1	(ALL).

P = People C = Committees

DATA FILE : P v C (1972/3)

KX(Y)

Value of q	Value of Q_q	Components
36	1	(SENATE).
30	2	(CHAIRS), (SENATE).
27	3	(CHAIRS), (SENATE), (COUNCIL).
26	4	(CHAIRS), (COMP-STUDS), (SENATE), (COUNCIL).
18	6	(CHAIRS), (SOCIAL-POL), (COMP-STUDS), (MATHS-STUDS), (SENATE), (COUNCIL).
16	8	(CHAIRS), (DISCIP), (SOCIAL-POL), (COMP-STUDS), (SOCIAL-STUDS), (MATHS-STUDS), (SENATE), (COUNCIL).
15	8	(CHAIRS, SENATE), (DISCIP), (SOCIAL-POL), (PHYS-SCI), (COMP-STUDS), (SOCIAL-STUDS), (MATHS-STUDS), (COUNCIL).
14	10	(CHAIRS, SENATE), (DISCIP), (ED-TECH), (SOCIAL-POL), (PHYS-SCI), (COMP-STUDS), (SOCIAL-STUDS), (MATHS-STUDS), (COUNCIL), (CAREERS).
11	10	(CHAIRS, SENATE, COUNCIL), (DISCIP), (ED-TECH), (SOCIAL-POL), (PHYS-SCI), (COMP-STUDS), (SOCIAL-STUDS), (MATHS-STUDS), (DEVELOPMENT), (CAREERS).
10	12	(CHAIRS, SENATE, COUNCIL), (DISCIP), (ED-TECH), (LIBRARY), (SOCIAL-POL), (PHYS-SCI), (COMP-STUDS), (SOCIAL-STUDS), (MATHS-STUDS), (DEVELOPMENT), (CAREERS), (APPEAL).
9	9	(CHAIRS, SENATE, COUNCIL, COMP-STUDS, SOCIAL-STUDS, MATHS-STUDS, DEVELOPMENT), (COMP), (DISCIP), (ED-TECH), (LIBRARY), (SOCIAL-POL), (PHYS-SCI), (CAREERS), (APPEAL).
8	13	(BURROWS), (CHAIRS, SENATE, COUNCIL, COMP-STUDS, SOCIAL-STUDS, MATHS-STUDS, DEVELOPMENT), (COMP), (DISCIP), (ED-TECH), (LIBRARY), (STAFFING), (SOCIAL-POL), (WIV-HOUSE), (PHYS-SCI), (FINANCE), (CAREERS), (APPEAL).
7	13	(BURROWS), (CARRERAS), (CHAIRS, SENATE, COUNCIL, STAFFING, COMP-STUDS, SOCIAL-STUDS, MATHS-STUDS, DEVELOPMENT, CAREERS), (COMP), (LATIN), (DISCIP), (ED-TECH), (LIBRARY), (SOCIAL-POL), (WIV-HOUSE), (PHYS-SCI), (FINANCE), (APPEAL).
6	11	(ACAD-PLAN, CHAIRS, SENATE, COUNCIL, STAFFING, MATHS-STUDS, BURROWS, COMP, COMP-STUDS, FINANCE, DEVELOPMENT, CAREERS), (ADMIS), (CARRERAS), (LATIN), (DISCIP), ((ED-TECH), (LIBRARY), (SOCIAL-POL), (WIV-HOUSE), (PHYS-SCI), (APPEAL).

P = People   C = Committees

DATA FILE : P v C (1972/3)(cont.)

KX(Y) (cont).

Value of q	Value of Q_q	Components
5	8	(ACAD-PLAN, CHAIRS, STAFFING, SENATE, COUNCIL, ADMIS, COMP-STUDS, SOCIAL-STUDS, MATHS-STUDS, BURROWS, COMP, LIBRARY, PHYS-SCI, FINANCE, DEVELOPMENT, CAREERS), (CARRERAS), (LATIN), (DISCIP), (ED-TECH), (SOCIAL-POL), (WIV-HOUSE), (APPEAL).
4	5	(ACAD-PLAN, ADMIS, CHAIRS, STAFFING, SENATE, COUNCIL, COMP-STUDS, SOCIAL-STUDS, MATHS-STUDS, DEVELOPMENT, BURROWS, CARRERAS, COMP, DISCIP, LIBRARY, SOCIAL-POL, PHYS-SCI, FINANCE, CAREERS, APPEAL), (LATIN), (ED-TECH), (BUXTON), (WIV-HOUSE).
0	1	(ALL).

P = People C = Committees

DATA FILE : C v DP

KY(X)

Value of q	Value of Q_q	Components
12	3	(CHAIRS), (ED-TECH), (SENATE).
11	2	(CHAIRS, SENATE), (ED-TECH).
10	1	(CHAIRS, SENATE, ED-TECH).
8	4	(CHAIRS, ED-TECH, SENATE), (LIBRARY), (COMP-STUDS), (SOCIAL-STUDS).
7	5	(BURROWS), (CHAIRS, ED-TECH, COMP-STUDS, SOCIAL-STUDS, MATHS-STUDS, SENATE, LIBRARY), (DISCIP), (SOCIAL-POL), (COUNCIL).
6	1	(BURROWS, CHAIRS, ED-TECH, LIBRARY, SENATE, DISCIP, STAFFING, SOCIAL-POL, COMP-STUDS, SOCIAL-STUDS, MATHS-STUDS, COUNCIL, CARRERAS, COMP).
5	1	(BURROWS, CHAIRS, ED-TECH, LIBRARY, COMP-STUDS, SOCIAL-STUDS, SENATE, COMP, DISCIP, STAFFING, SOCIAL-POL, MATHS-STUDS, COUNCIL, CARRERAS, LATIN, WIV-HOUSE, DEVELOPMENT, CAREERS).
4	1	(ACAD-PLAN, CHAIRS, ED-TECH, SENATE, COUNCIL, BURROWS, CARRERAS, COMP, LATIN, DISCIP, LIBRARY, STAFFING, SOCIAL-POL, PHYS-SCI, COMP-STUDS, SOCIAL-STUDS, MATHS-STUDS, BUXTON, WIV-HOUSE, DEVELOPMENT, CAREERS).
3	1	(ACAD-PLAN, ADMIS, CARRERAS, CHAIRS, ED-TECH, STAFFING, MATHS-STUDS, SENATE, COUNCIL, DEVELOPMENT, BURROWS, BUXTON, SOCIAL-POL, COMP-STUDS, SOCIAL-STUDS, COMP, LATIN, DISCIP, LIBRARY, WIV-HOUSE, PHYS-SCI, CAREERS, FINANCE, APPEAL).
2	1	(ACAD-PLAN, ADMIS, BURROWS, CARRERAS, CHAIRS, COMP, DISCIP, ED-TECH, STAFFING, PHYS-SCI, SOCIAL-STUDS, MATHS-STUDS, SENATE, COUNCIL, FINANCE, DEVELOPMENT, DISC-APPEAL, LATIN, LIBRARY, MEMBERSHIP, SOCIAL-POL, WIV-HOUSE, COMP-STUDS, CAREERS, APPEAL, BUXTON, PENGUIN).
1	1	(ACAD-PLAN, ADMIS, BURROWS, CARRERAS, CHAIRS, COMP, LATIN, DISCIP, ED-TECH, LIBRARY, BUXTON, STAFFING, SOCIAL-POL, WIV-HOUSE, PHYS-SCI, COMP-STUDS, SOCIAL-STUDS, MATHS-STUDS, SENATE, COUNCIL, FINANCE, DEVELOPMENT, APPEAL, DISC-APPEAL, MEMBERSHIP, CAREERS, PENGUIN, MEM-REVIEW).

C = Committees   DP = Departments

DATA FILE : C v DP

KX(Y)

Value of q	Value of Q_q	Components
20	1	(AD).
15	3	(AD), (EE), (PH).
14	5	(AD), (MA), (ECON), (EE), (PH).
13	8	(AD), (MA), (GOV), (ECON), (CC), (EE), (PH), (ST).
12	7	(AD, MA), (GOV), (ECON), (CC), (EE), (PH), (ST).
11	6	(AD, MA, ECON, PH), (GOV), (CC), (LIT), (EE), (ST).
10	4	(AD, MA, GOV, ECON, CC, EE, PH), (SOC), (LIT), (ST).
9	2	(AD, MA, GOV, ECON, SOC, CC, LIT, EE, PH, ST), (LA).
8	2	(AD, MA, GOV, ECON, SOC, CC, LIT, EE, HIST, PH, LAY, ST), (LA).
7	1	(AD, MA, GOV, ECON, SOC, CC, LIT, EE, HIST, PH, LAY, ST, LA).
5	2	(AD, MA, GOV, ECON, SOC, CC, LIT, LA, EE, HIST, PH, LAY, ST), (AR).
4	1	(AD, MA, GOV, ECON, SOC, CC,LLIT, LA, EE, HIST, PH, LAY, AR, ST, LIB).
1	1	(AD, MA, GOV, ECON, SOC, CC, LIT, LA, EE, HIST, PH, LAY, AR, ST, LIB, CH).

C = Committees   DP = Departments

DATA FILE : C v DP (SV = 3)

KY(X)

Value of q	Value of Q_q	Components
7	1	(SENATE).
5	2	(COMP-STUDS), SENATE).
3	2	(CHAIRS, SENATE), (COMP-STUDS).
2	3	(CHAIRS, SENATE, PHYS-SCI, COMP-STUDS, MATHS-STUDS), (SOCIAL-STUDS), (COUNCIL).
1	3	(CHAIRS, COMP-STUDS, SOCIAL-STUDS, MATHS-STUDS, SENATE, PHYS-SCI, COUNCIL), (DISCIP), (CAREERS).
0	1	(ACAD-PLAN, ADMIS, CHAIRS, MATHS-STUDS, SENATE, COUNCIL, ED-TECH, COMP-STUDS, SOCIAL-STUDS, DISCIP, PHYS-SCI, FINANCE, DEVELOPMENT, CAREERS, APPEAL, SOCIAL-POL, WIV-HOUSE).

KX(Y)

5	2	(MA), (GOV).
4	4	(MA), (GOV), (LAY), (ST).
3	5	(MA), (GOV),(CC), (LAY), (ST).
2	5	(MA, GOV, ECON, LIT), (CC), (PH), (LAY), (ST).
1	3	(MA, GOV, ECON, CC, LIT, PH, SOC, LA, EE), (ST).
0	1	(MA, GOV, ECON, CC, LIT, LA, EE, PH, LAY, SOC, HIST, ST).

C = Committees DP = Departments

DATA FILE: C v B1 ((N+1)-level)

KY(X)

Value of q	Value of Q_q	Components
14	1	(3CTTEE).
12	3	(3CTTEE), (4CTTEE), (7CTTEE).
11	4	(3CTTEE), (4CTTEE), (7CTTEE), (8CTTEE, 9CTTEE, 10CTTEE, 11CTTEE).
9	4	(3CTTEE), (4CTTEE, 5CTTEE), (7CTTEE), (8CTTEE, 9CTTEE, 10CTTEE, 11CTTEE).
7	5	(1CTTEE), (3CTTEE), (4CTTEE, 5CTTEE), (7CTTEE), (8CTTEE, 9CTTEE, 10CTTEE, 11CTTEE).
6	4	(1CTTEE, 2CTTEE), (3CTTEE), (4CTTEE, 5CTTEE, 7CTTEE), (8CTTEE, 9CTTEE, 10CTTEE, 11CTTEE).
3	3	(1CTTEE, 2CTTEE), (3CTTEE, 7CTTEE, 4CTTEE, 5CTTEE), (8CTTEE, 9CTTEE, 10CTTEE, 11CTTEE).
0	2	(1CTTEE, 2CTTEE, 4CTTEE, 5CTTEE, 7CTTEE, 3CTTEE), (8CTTEE, 9CTTEE, 10CTTEE, 11CTTEE).

KX(Y)

Value of q	Value of Q_q	Components
3	1	(37BM1, 38BM1, 39BM1, 40BM1, 41BM1, 43BM1, 44BM1, 45BM1, 46BM1, 47BM1, 48BM1, 49BM1).
2	3	(38BM1), (14BM1, 16BM1, 18BM1, 19BM1, 20BM1, 21BM1), (37BM1, 38BM1, 39BM1, 40BM1, 41BM1, 43BM1, 44BM1, 45BM1, 46BM1, 47BM1, 48BM1, 49BM1).
1	4	(2BM1, 3BM1, 4BM1, 5BM1, 6BM1, 7BM1, 8BM1), (9BM1, 14BM1, 16BM1, 18BM1, 19BM1, 20BM1, 21BM1, 12BM1, 13BM1, 15BM1, 17BM1), (22BM1, 30BM1, 33BM1, 36BM1), (37BM1, 38BM1, 39BM1, 40BM1, 41BM1, 43BM1, 44BM1, 45BM1, 46BM1, 47BM1, 48BM1, 49BM1).

DATA FILE : C v B1 ((N+1)-level)

KX(Y) (cont).

Value of q	Value of Q_q	Components
0	3	(1BM1, 2BM1, 3BM1, 4BM1, 5BM1, 6BM1, 7BM1, 8BM1, 9BM1, 12BM1, 13BM1, 14BM1, 15BM1, 16BM1, 17BM1, 18BM1, 19BM1, 20BM1, 21BM1, 42BM1, 10BM1, 11BM1, 22BM1, 30BM1, 33BM1, 36BM1, 23BM1, 24BM1, 25BM1, 26BM1, 27BM1, 28BM1, 29BM1, 31BM1, 32BM1, 34BM1, 25BM1), (37BM1, 38BM1, 39BM1, 40BM1, 41BM1, 43BM1, 44BM1, 45BM1, 46BM1, 47BM1, 48BM1, 49BM1).

C = Committees B1 = Business Items

DATA FILE : C v B (N-level)

KY(X)

Value of q	Value of Q_q	Components
14	2	(8CTTEE), (9CTTEE).
12	3	(8CTTEE, 9CTTEE), (10CTTEE), (11CTTEE).
11	1	(8CTTEE, 9CTTEE, 10CTTEE, 11CTTEE).
10	2	(6CTTEE), (8CTTEE, 9CTTEE, 10CTTEE, 11CTTEE).
9	4	(4CTTEE), (6CTTEE), (7CTTEE), (8CTTEE, 9CTTEE, 10CTTEE, 11CTTEE).
8	2	(4CTTEE), (6CTTEE, 8CTTEE, 9CTTEE, 10CTTEE, 11CTTEE, 7CTTEE)
3	4	(1CTTEE), (4CTTEE), (6CTTEE, 8CTTEE, 9CTTEE, 10CTTEE, 11CTTEE), (7CTTEE).
2	4	(1CTTEE, 7CTTEE), (3CTTEE), (4CTTEE), (6CTTEE, 8CTTEE, 9CTTEE, 10CTTEE, 11CTTEE).
1	2	(1CTTEE, 2CTTEE, 7CTTEE, 3CTTEE, 4CTTEE), (6CTTEE, 8CTTEE, 9CTTEE, 10CTTEE, 11CTTEE).
0	1	(1CTTEE, 2CTTEE, 3CTTEE, 4CTTEE, 6CTTEE, 7CTTEE, 8CTTEE, 9CTTEE, 10CTTEE, 11CTTEE, 5CTTEE).

C = Committees B = Business Items

DATA FILE : C v B (N-level)

KX(Y)

Value of q	Value of Q_q	Components
5	1	(300BSNS).
4	1	(10BSNS, 20BSNS, 30BSNS, 50BSNS, 29BSNS, 300BSNS).
3	3	(10BSNS, 20BSNS, 30BSNS, 40BSNS, 50BSNS, 29BSNS, 300BSNS, 310BSNS, 340BSNS, 350BSNS, 380BSNS), (190BSNS), (210BSNS).
2	3	(10BSNS, 20BSNS, 30BSNS, 40BSNS, 50BSNS, 100BSNS, 290BSNS, 300BSNS, 310BSNS, 340BSNS, 350BSNS, 370BSNS, 380BSNS), (190BSNS, 210BSNS), (260BSNS).
1	2	(10BSNS, 20BSNS, 30BSNS, 40BSNS, 50BSNS, 70BSNS, 90BSNS, 100BSNS, 290BSNS, 300BSNS, 310BSNS, 320BSNS, 340BSNS, 350BSNS, 360BSNS, 370BSNS, 380BSNS), (190BSNS, 200BSNS, 210BSNS, 220BSNS, 260BSNS).
0	1	(10BSNS, 20BSNS, 30BSNS, 40BSNS, 50BSNS, 70BSNS, 90BSNS, 100BSNS, 290BSNS, 300BSNS, 310BSNS, 320BSNS, 330BSNS, 340BSNS, 350BSNS, 360BSNS, 370BSNS, 380 BSNS, 410BSNS, 190BSNS, 200BSNS, 210BSNS, 220BSNS, 230BSNS, 240BSNS, 260BSNS, 270BSNS, 280BSNS, 390BSNS, 420BSNS, 110BSNS, 120BSNS, 130BSNS, 140BSNS, 150BSNS, 160BSNS, 170BSNS, 180BSNS).

C = Committees B = Business Items

Index

Interdisciplinary Systems Research
Birkhäuser Verlag, Basel und Stuttgart

ISR 1
René Hirsig:
Menschliches Konformitätsverhalten – am Computer simuliert

ISR 2
Werner Hugger:
Weltmodelle auf dem Prüfstand

ISR 3
Claus Schönebeck:
Der Beitrag komplexer Stadtsimulationsmodelle (vom Forrester-Typ) zur Analyse und Prognose großstädtischer Systeme

ISR 4
Christof W. Burckhardt (Editor):
Industrial Robots – Robots industriels – Industrieroboter

ISR 5
Kuno Egle:
Entscheidungstheorie

ISR 6
Dieter Ruloff:
Konfliktlösung durch Vermittlung: Computersimulation zwischenstaatlicher Krisen

ISR 7
Salomon Klaczko:
Systemanalyse der Selbstreflexion

ISR 8
John Craig Comfort:
A Flexible Efficient Computer System to Answer Human Questions

ISR 9
Richard Rickenbacher:
Lernen und Motivation als relevanzgesteuerte Datenverarbeitung

ISR 10
James Gips:
Shape Grammars and their Uses

ISR 11
C. V. Negoita / D. A. Ralescu:
Applications of Fuzzy Sets to System Analysis

ISR 12
Jean Vuillemin:
Syntaxe, sémantique et axiomatique d'un langage de programmation simple

ISR 13
George Stiny:
Pictorial and Formal Aspects of Shape, Shape Grammars and Aesthetic Systems

ISR 14
Alexander Van der Bellen:
Mathematische Auswahlfunktionen und gesellschaftliche Entscheidungen

ISR 15
Dana H. Ballard:
Hierarchic Recognition of Tumors in Chest Radiographs

ISR 16
James R. Low:
Automatic Coding: Choice of Data Structures

ISR 17
Richard Young:
Seriation by Children: An Artificial Intelligence Analysis of a Piagetian Task

ISR 18
Helmut Maier:
Computersimulation mit dem Dialogverfahren SIMA
Bd. 1: *Konzeption*

ISR 19
Bd. 2: *Dokumentation*

ISR 20
Hartmut Bossel / Salomon Klaczko / Norbert Müller (Editors):
System Theory in the Social Sciences

ISR 21
Ekkehard Brunn / Gerhard Fehl (Hrsg.):
Systemtheorie und Systemtechnik in der Raumplanung

ISR 22
Remakant Nevatia:
Computer Analysis of Scenes of 3-dimensional Curved Objects

ISR 23
Henry M. Davis:
Computer Representation of the Stereochemistry of Organic Molecules

ISR 24
Bernhelm Booss / Klaus Krickeberg (Hrsg.):
Mathematisierung der Einzelwissenschaften

ISR 25
Claus W. Gerberich:
Alternativen in der Forschungs- und Entwicklungspolitik eines Unternehmens

ISR 26
Hans-Paul Schwefel:
Numerische Optimierung von Computer-Modellen mittels der Evolutionsstrategie

ISR 27
Hermann Krallmann:
Heuristische Optimierung von Simulationsmodellen mit dem Razor-Search Algorithmus

ISR 28
Stefan Rath-Nagel:
Alternative Entwicklungsmöglichkeiten der Energiewirtschaft in der BRD

ISR 29
Harry Wechsler:
Automatic Detection of Rib Contours in Chest Radiographs

ISR 30
Alfred Voss:
Ansätze zur Gesamtanalyse des Systems Mensch–Energie–Umwelt

ISR 31
Dieter Eberle:
Ein Computermodell der Verflechtung zwischen Wohn- und Naherholungsgebieten der Region Hannover

ISR 32
Ernst Billeter / Michel Cuénod / Salomon Klaczko:
Overlapping Tendencies in Operations Research, Systems Theory and Cybernetics

ISR 33
G. Matthew Bonham / Michael J. Shapiro (Editors)
Thought and Action in Foreign Policy

ISR 34
Ronald H. Atkin
Combinatorial Connectivities in Social Systems

ISR 35
Moscheh Mresse
MOSIM – ein Simulationskonzept basierend auf PL/I

ISR 36/37/38
Hartmut Bossel (Editor)
Concepts and Tools of Computer-Assisted Policy Analysis
Volume 1: Basic Concepts
Volume 2: Causal Systems Analysis
Volume 3: Cognitive Systems Analysis

Interdisziplinäre Systemforschung

MIX
Papier aus verantwortungsvollen Quellen
Paper from responsible sources
FSC® C105338

If you have any concerns about our products,
you can contact us on
ProductSafety@springernature.com

In case Publisher is established outside the EU,
the EU authorized representative is:
Springer Nature Customer Service Center GmbH
Europaplatz 3, 69115 Heidelberg, Germany

Printed by Libri Plureos GmbH
in Hamburg, Germany